Fundamentals of Spatial Data Quality

Fundamentals of Spatial Data Quality

Edited by
Rodolphe Devillers
Robert Jeansoulin

Part of this book adapted from "Qualité de l'information géographique" published in France by Hermès Science/Lavoisier in 2005
Published in Great Britain and the United States in 2006 by ISTE Ltd

ISTE Ltd   ISTE USA
6 Fitzroy Square   4308 Patrice Road
London W1T 5DX   Newport Beach, CA 92663
UK   USA

www.iste.co.uk

Library of Congress Cataloging-in-Publication Data

Fundamentals of spatial data quality / edited by Rodolphe Devillers, Robert Jeansoulin.
    p. cm.
  "Part of this book adapted from "Qualité de l'information géographique"
published in France by Hermes Science/Lavoisier in 2005."
  Includes bibliographical references and index.
  ISBN-13: 978-1-905209-56-9
  ISBN-10: 1-905209-56-8
  1.   Geographic information systems--Data processing--Quality control.   I. Devillers, Rodolphe. II. Jeansoulin, Robert.
  G70.212.F85 2006
  526.0285--dc22

                                                                    2006009298

British Library Cataloguing-in-Publication Data
A CIP record for this book is available from the British Library
ISBN 10: 1-905209-56-8
ISBN 13: 978-1-905209-56-9

Table of Contents

Chapter 9. Spatial Integrity Constraints: A Tool for Improving the Internal Quality of Spatial Data 161
Sylvain VALLIERES, Jean BRODEUR and Daniel PILON

Chapter 10. Quality Components, Standards, and Metadata 179
Sylvie SERVIGNE, Nicolas LESAGE and Thérèse LIBOUREL

Foreword

The quality of spatial data, as indeed of any data, is crucial to its effective use. Spatial data purport to represent aspects of the spatial world, and in the context of this book that means primarily the geographic world, the world defined by human experience and by the surface and near-surface of the Earth. Quality, as the definitions in this book attest, is a measure of the difference between the data and the reality that they represent, and becomes poorer as the data and the corresponding reality diverge. Thus, if data are of poor quality, and tell us little about the geographic world, then they have little value.

This argument seems watertight, and the examples in the Introduction provide ample illustration. But as Nicholas Chrisman notes in Chapter 1, the initial reaction of many leaders of the field was negative. My own experience was similar; in 1977, I gave a presentation on data quality to an international conference of experts in spatial data, and was met with something between indifference and outright opposition. In the almost 30 years since then, there has been no massive outcry among users of geographic information systems (GIS) and spatial databases, demanding better methods of handling the uncertainty present in data; despite many warnings, there have been few court cases over bad decisions that resulted from poor data, and the major GIS vendors still provide little in the way of support for handing information about data quality. Yet the geographic information science (GIScience) community continues to identify data quality as a topic of major significance, and much progress continues to be made, as the chapters of this book will confirm.

I think that there are several explanations for this apparent contradiction, and they lie at the very heart of GIScience. First, these issues cannot be ignored by anyone with a scientific conscience. It makes no sense whatsoever for a GIS vendor to claim that his or her software stores coordinates to double precision, in other words to 14 significant digits, when one part in 10^{14} of the linear dimension of the Earth is approximately the size of a molecule. None of our devices for measuring position have accuracies that are any better than one part in 10^7, or single precision, but in this and in many other instances it is the precision of the digital computer that masquerades as accuracy. Any self-respecting scientist knows that it is misleading to report any result to a precision that exceeds its accuracy, yet our GIS software does so constantly. It is clearly the responsibility of the GIScience community to draw attention to such issues, to reflect on the role that software plays, and to demand that it adhere to the best of scientific principles.

Second, there is a long tradition in map-making of compromising the objective of portraying the world accurately with the potentially conflicting objective of visual clarity. A contour might be kinked, for example, to emphasize the presence of a stream, whether or not the stream's course is actually indented in the landscape. A railway running close to a road might be separated from it in the interests of avoiding visual confusion. Thus a map can be far from a scientifically accurate representation of the Earth's surface, yet it is natural to assume that the contents of a map, once digitized, stored in a database, and analyzed using GIS software, are indeed scientifically accurate. The cartographic perspective even leads to a somewhat different interpretation of data quality – a digitized map can be said to be perfect if it exactly represents the contents of the paper map from which it was obtained, whether or not the paper map exactly represents phenomena in the real world.

Third, while many of the methods discussed in this and other books on the topic of spatial data quality are intuitive and simple, the theoretical frameworks in which they are grounded – spatial statistics, geostatistics, and set theory – are complex and difficult. Quality is difficult to attach to individual features in a database, but instead must be described in terms of the joint quality of pairs of features, through measures of relative positional accuracy, covariance or correlation. Surveyors have dealt with these problems for decades, and have developed appropriate training regimes for their students, but many users of GIS lack the necessary mathematical skills to handle complex models of spatial data quality. Instead, researchers have had to look for clever visual ways of capturing and communicating what is known about quality, about how quality varies from one type of feature to another, and about how it varies from one geographic area to another. And as with any technology that makes difficult mathematical concepts accessible to a broad community of users, there is always the potential for misinterpretation and misuse.

These three issues are the common threads that have run through the work that I have done on the topic of spatial data quality over the past three decades. Thinking back, my own interest in the topic seems to have stemmed from several intersecting themes and ideas. First, I was fascinated with the field of geometric probability, and the elegant results that had been obtained by such mathematicians as Buffon and Coxeter – and thought that these ideas could be applied to maps. As Ashton

Shortridge and I showed in a paper many years later, the statistics of a vector crossing a raster cell can be related to Buffon's famous problem of the needle randomly dropped on a set of parallel lines ([SHO 02]). Second, I was bothered by the lack of any simple models of the errors introduced by digitizing, or of the uncertainties inherent in such simple GIS operations as the measurement of area. One of the paradoxes of GIS is that it is possible to estimate properties such as slope accurately even from a very inaccurate digital elevation model, because slope responds to the covariance of errors in addition to the variance, as many GIScientists have shown (see, for example, [HUN 97]). Third, I was struck by the wealth of knowledge in disciplines such as geostatistics and surveying that was virtually unknown to the GIScience community. A paper on measurement-based GIS ([GOO 02]), for example, was prompted by what I perceived as a need to bring research on adjustment theory into the GIScience literature.

Much of the early work in GIS was dominated by the desire to create accurate digital representations of the contents of maps. The Canada Geographic Information System of the 1960s, for example, saw as its primary mission the capture of mapped information on land, followed by calculation and tabulation of area; many other early GIS projects had similar goals. Much later, GIScientists began to look systematically at the results of these projects, and the degree to which their results replicated not the contents of maps, but the contents of the real world. By that time, of course, many of the fundamental design decisions of GIS had been made. Those decisions were predicated on the assumption that it was possible to create a perfect representation of the contents of a map, and even today that assumption seems reasonable. But, as I am sure all of the authors of the chapters of this book would argue, it is not possible to create a perfect representation of the infinite complexity of the real world. If the field of GIS had begun in the mid-1960s with that assumption, one might reasonably ask whether the design decisions would have been the same. Does a technology designed for the goal of perfect representation of the contents of maps adapt well to the imperfect representation of the contents of the real world? This seems to me to be one of the most profound questions that GIScience can ask – in effect, it asks whether the ontological legacy of GIS is consistent with its fundamental objectives.

The chapters of this book present an excellent overview of the dimensions of spatial data quality research, from the most theoretical and abstract to the most practical and applied. The book has no epilog or concluding chapter, so perhaps I might be permitted to offer a few comments on where the field might be headed. Previous comments notwithstanding, there does appear to be steady progress in the adoption of a greater sensitivity to spatial data quality issues among the user community and GIS software vendors. Better standards for description of spatial data quality are being adopted, and are being supported by software. Suppliers of data are more likely to provide statements of data quality, and to test products against ground truth, than they were in the past. A greater range of examples is available in the literature, and spatial data quality is now an obligatory part of the GIS curriculum. This book, and its availability in English, will add substantially to that literature.

That said, however, the central problem seems as unsolved as ever – how to communicate what is known about spatial data quality to an ever-expanding population of users, many of whom have very little understanding of the basic principles of GIScience. In the past year, we have seen a massive expansion of access to spatial data, through the introduction and widespread popularity of Google Earth and similar tools. Very few of the people recruited to the use of spatial data by these technologies will have any understanding of spatial data quality issues, but many of them will likely have the motivation to learn, if the research community can develop and implement appropriate techniques. This book and its coverage of the important issues should keep us moving in the right direction.

Michael F. Goodchild

Bibliography

[GOO 02] M.F. GOODCHILD (2002) "Measurement-based GIS" in W. SHI, P.F. FISHER and M.F. GOODCHILD, editors, *Spatial Data Quality*. New York: Taylor & Francis, pp. 5–17.

[HUN 97] G.J. HUNTER and M.F. GOODCHILD (1997) "Modeling the uncertainty in slope and aspect estimates derived from spatial databases", *Geographical Analysis* 29(1): 35–49.

[SHO 02] A.M. SHORTRIDGE and M.F. GOODCHILD (2002) "Geometric probability and GIS: some applications for the statistics of intersections", *International Journal of Geographical Information Science* 16(3): 227–243.

Introduction

The quality of geographic information (or geospatial data quality) has always presented a significant problem in geomatics. It has experienced significant growth in recent years with the development of the World Wide Web, increased accessibility of geomatic data and systems, as well as the use of geographic data in digital format for various applications by many other fields.

Problems regarding data quality affect all fields that use geographic data. For example, an environmental engineer may need to use a digital elevation model to create a model of a watershed, a land surveyor may need to combine various data to obtain an accurate measurement of a given location, or someone may simply need to surf the Web to find an address from an online map site. Although many people associate the term "quality" only with the spatial accuracy of collected data (e.g. data collected by Global Positioning System (GPS) in relative mode will be of "better quality" than those digitized from paper maps at a scale of 1:100,000), the concept of quality encompasses a much larger spectrum and affects the entire process of the acquisition, management, communication, and use of geographic data.

Research into certain aspects of geographic data quality has been ongoing for a number of years, but interest in the subject has been increasing more recently. In the 1990s, the scientific community began to hold two conferences on this subject:

– *Accuracy* (*International Symposium on Spatial Accuracy Assessment in Natural Resources & Environmental Sciences*): addresses uncertainty regarding, primarily, the field of natural resources and the environment. This conference has been held every two years for the past ten years.

– *ISSDQ* (*International Symposium on Spatial Data Quality*): addresses geographic data quality in general. This conference has been held every two years since 1999.

Written by Rodolphe DEVILLERS and Robert JEANSOULIN.

In addition to events dealing entirely with spatial data quality, many large conferences on geomatics and on other related fields hold sessions on spatial data quality. Several international organizations also have working groups that address issues regarding quality, such as the International Society for Photogrammetry and Remote Sensing (ISPRS) (WG II/7: *Quality of Spatio-Temporal Data and Models*), the Association of Geographic Information Laboratories in Europe (AGILE) (*WG on Spatial Data Usability*), and the International Cartographic Association (ICA) (*WG on Spatial Data Uncertainty and Map Quality*). Quality is also a concern for standardization bodies that address it in a general way for the production and distribution of goods and services (for example, ISO 9000), but also for the field of geographic information specifically, often with regard to the standardization of metadata (for example, FGDC, OGC, CEN, ISO TC/211).

Several books have been published over the last decade on the subject of quality [GUP 95; GOO 98; SHI 02]. The first [GUP 95] presents reflections on the elements of quality by members of the *Spatial Data Quality* commission of the International Cartographic Association. The two other books [GOO 98 and SHI 02] present research breakthroughs on issues regarding quality and the uncertainty of geographic information. Other books have also been published about the problems of uncertainty in geographic data (for example [ZHA 02; FOO 03]). The present book differs from the above-mentioned publications by attempting to offer a more global, complete, and accessible vision of quality, intended for a wider range of users and not only for experts in the field.

Decision and uncertainty: notion of usefulness of quality

The notion of usefulness of quality depends first on the use that is made of the data it quantifies, particularly during and after a decision. Decision and quality are part of a dialectic that is similar to that of risk and hazard. Decision may be regarded as the conclusion of an informed and logical process, in which the treatment of uncertainty[1] must be present. The purpose of the usefulness of quality is then measured by its ability to reduce the uncertainty of a decision. Research into probability and statistics, artificial intelligence, and databases have examined the question of uncertainty for many years. Spatial and temporal data are largely present today: for example, in conferences such as UAI (*Uncertainty and Artificial Intelligence*), in international conferences on Information Processing and Management of Uncertainty in Knowledge-Based Systems (IPMU), and in European Conferences on Symbolic and Quantitative Approaches to Reasoning and Uncertainty (ECSQARU).

---

1 The concept of uncertainty is presented in more detail in Chapter 3.

The purpose of this book is to present a general view of geographic data quality, through chapters that combine basic concepts and more advanced subjects. This book contains 15 chapters grouped into four parts and includes an appendix.

Organization of the work

The first part, entitled "Quality and Uncertainty: Introduction to the Problem", introduces the work and provides certain basic concepts. Chapter 1 presents a brief history of the study of geographic data quality, showing how the field developed and grew from strict problems of spatial accuracy to include much more comprehensive considerations, such as the assessment of fitness for use. This historical perspective allows a better understanding of how certain aspects of the study of quality developed to become what it is today. Chapter 2 presents general concepts on quality, definitions, certain sources of the problem with quality, as well as the fundamental distinction in geographic information between what is called "internal quality" and "external quality". Chapter 3 positions and defines different terms found in the field of data quality and uncertainty (for example, uncertainty, error, accuracy, fuzziness, vagueness) and describes these different types of uncertainties using examples.

The second part is titled "Academic Case Studies: Raster, Choropleth and Land Use". Data quality is often determined at the moment of data acquisition (for example, depending on the technique used). Chapter 4 presents problems associated with quality in the field of imagery, taking into account geometric and radiometric factors. Chapter 5 addresses the inaccuracy of maps as a result of misinterpretation, highlighting the difficulty of classifying natural environments in which precise boundaries between phenomena do not always exist. Chapter 6 presents statistical methods used to measure and manipulate uncertainties related to data. Finally, Chapter 7 presents several methods of quantitative and qualitative reasoning that allow the manipulation of uncertain data. These concepts are illustrated through an example of land use.

The third part, entitled "Internal Quality of Vector Data: Production, Evaluation and Documentation", presents the problems of quality at different stages of producing vector data. Chapter 8 addresses the concerns of a geographic vector data producer for quality related issues (for example, quality control and quality assurance). Chapter 9 presents a specific approach to improve internal data quality during production, based on the definition of rules that verify whether the objects have possible relationships according to their semantic. Chapter 10 presents how to describe the quality of data produced, as well as national and international standards on the description of quality and its documentation in the form of metadata. Finally, Chapter 11 complements Chapter 10, by presenting an evaluation of quality from a conceptual point of view and then addressing several methods that can be used to measure quality.

The fourth and final part, entitled "External Quality: Communication and Usage", takes the point of view of the user, who bases his or her decisions on the data. Chapter 12 addresses the communication of information on quality to users and the use of information by users. It discusses questions such as the management and visual representation of quality information. Chapter 13 presents a formal approach based on ontologies to evaluate the fitness of use of data for a given purpose (that is, external quality). Chapter 14 addresses the relationship between data quality and the decision-making process based on these data. Finally, Chapter 15 deals with legal considerations, such as civil liability associated with data quality.

In conclusion, we hope that this book will help to answer many questions regularly asked regarding the quality of geographic information:

− Everyone says that data quality is important; is that true? Yes. It is essential. Failure to consider data quality may result in serious consequences.

− Everyone says that quality is complex? Yes, but there are a growing number of methods to overcome this complexity.

− Is there a better way to consider quality? No. It is impossible to answer, *a priori* and in a unique way, the question of whether or not the data are good, even in the case of a precise application defined in advance. Solutions exist however, that clarify the use made of data, for a given purpose, according to known information on their quality. The world is not deterministic. The ultimate decision is subjective and depends on the user.

Bibliography

[FOO 03] FOODIE G.M. and ATKINSON P.M. (eds), *Uncertainty in Remote Sensing and GIS*, 2003, New York, John Wiley & Sons, p 326.

[GOO 98] GOODCHILD M. and JEANSOULIN R. (eds), *Data Quality in Geographic Information: From Error to Uncertainty*, 1998, Paris, Hermès Science Publications.

[GUP 95] GUPTILL S.C. and MORRISON J.L. (eds), *Elements of Spatial Data Quality*, 1995, Oxford, UK, Pergamon Press, p 250.

[SHI 02] SHI W., GOODCHILD M.F. and FISHER P.F. (eds), *Spatial Data Quality*, 2002, London, Taylor and Francis, p 336.

[ZHA 02] ZHANG J. and GOODCHILD M.F., *Uncertainty in Geographical Information*, 2002, London, Taylor and Francis, p 192.

Chapter 1

Development in the Treatment
of Spatial Data Quality

1.1. Introduction

Twenty-three years ago, I traveled to Ottawa to present a paper entitled: "The role of quality information in the long-term functioning of a geographic information system (GIS)". Issues of data quality were not prominent at AUTO-CARTO 6, an international event. My presentation was scheduled at the very end of a long day when most participants were heading off for more relaxing pursuits. Certain well-established GIS experts (including both Roger Tomlinson and Duane Marble) doubted my thesis that quality information would take up as much space in data storage as geometric coordinates. More importantly, my paper argued that conceptual work had ignored the issue. Now, after more than 20 years, it should be abundantly evident that data quality issues have risen in prominence. It is no longer possible to ignore the role of quality information in studies of geographic information. On the conceptual front, there have been some advances as well.

To introduce this book on recent research, it is important to review the development of concepts of data quality for geographic information. This chapter will explore the boundaries of studies of data quality, covering the philosophical as much as the technical matters.

Chapter written by Nicholas CHRISMAN.

1.2. In the beginning

Data quality is a complex concept that has evolved from many sources over a long period of time. In the field of cartography, the term has become important, but the use of this overarching concept is fairly recent. The concern with quality is linked to the attention paid in general business management to quality concerns in the so-called "quality circles" of Swedish manufacturing and other such initiatives. Before that attention, the more operative term was "accuracy". For example, the US National Map Accuracy Standards (NMAS) [BUR 47] deal exclusively with positional accuracy, and only 'well defined' points – those objects that are more easily tested.

NMAS was developed in the 1940s by professional photogrammetrists for a very practical purpose. It specified a specific threshold – 0.5 mm or 0.8 mm on the printed map – depending on the scale. If the 90% of the points tested fell within this threshold, the map complied and could bear the statement "This map complies with NMAS". It is somewhat curious in our modern era of openness that a map that failed the test would simply omit any mention of the standard. While NMAS was specifically national, other countries adopted approaches that were similar to or simply copied from NMAS.

NMAS fits into a general approach to quality where a single threshold serves as the standard expectation [HAY 77]. Quality consists of conforming to this threshold. Clearly, this strategy works best under static technology where the threshold reflects what can be expected from available techniques. It also works best for a stable base of users whose needs are served by this particular threshold. This arrangement describes the engineering uses of topographic maps in the pre-digital era. It is important to recognize that there is a tacit correspondence between the technically possible, the economically viable, and the utility for the intended uses. Without that connection, the fixed-threshold test would be pointless.

One of the first developments in conceptual understanding of data quality involved the adoption of statistical concepts. While standard deviation and simple tests of distribution have become entirely routine, they were not in common use in the 1940s when NMAS was adopted. There were a number of distinct threads that brought operations research and quantification with statistics into many sectors of society. In the matter of measures of map accuracy, the bomb damage assessment studies immediately following World War II led directly into the missile accuracy questions of the Cold War [MAC 90; CLO 02]. The difference was that the distribution of error was treated with statistical tools in the case of missile accuracy. Instead of specifying that 90% of the points tested must stay inside a given threshold, the "Circular Map Accuracy Standard" (CMAS) promulgated by the military ACIC used the standard deviation to calculate the distance in the normal distribution that would contain 90% of the predicted population. Of course, this converted a simple test, which one either passed or failed, into a statement about uncertainty. Depending on the distribution of the other points, it would be possible to fail a CMAS test, but pass under NMAS or vice versa. It took until 1989 for the statistical approach of CMAS to be formally adopted by the photogrammetric organization that originated NMAS [ASP 89], and CMAS is now the basis for the replacement for NMAS adopted by US Geological Survey and the Federal Geographic Data Committee, the National Standard for Spatial Data Accuracy [FGD 98].

NMAS also contains some rather problematic concepts when it comes to testing. The positions on the map are intended to be compared to the 'true' positions, without much consideration about how the truth is going to be approximated. Certainly, between 1940s and 1960s, the practice of frequentist statistics asserted that the central tendency of many measurements revealed the underlying true value. Later in this chapter, I will deal with alternative approaches to locating the truth that are more defensible.

1.3. Changing the scene

The data quality scene, originally focused on positional accuracy, developed in a number of directions to broaden the topic and to change the basic approach to information management. Space constraints mean that all these changes cannot be examined, but I will consider three changes with large ramifications. First, I will deal with the recognition of attribute accuracy in its origins in remote sensing. Next, I will deal with the topological movement and its role in practical issues in the verification of logical consistency. Finally, I will cover the emergence of the fitness for use concept.

1.3.1. *Accuracy beyond position*

From its origins in positional accuracy assessment, the idea of map accuracy had to be broadened to deal with the full information content. This means addressing the 'attribute' of a thematic map, the object identities of an inventory on an engineering drawing, and whatever else a user will question. In an era when numbers were given a higher value than 'mere' categories, this hardly looked like a profitable field of research, but the issue became important in a number of practical settings. One of the most crucial ones involved the assessment of success in classifying remotely-sensed images. In the visual world of photointerpretation, the identity of objects was controlled by keys and indexes to images where the "ground truth" was known. When the first digital images were provided from satellites, they were treated as visual material in much the same way. NASA funded many applications projects using the early remote sensing capability, and the research community responded with estimates of "percentage correctly classified", often rather dubiously overestimated using some less-than-independent methodologies. Inside the remote sensing profession, some questioned the utility of the single measure of the percentage in the diagonal of the misclassification matrix [CON 91]. Some statistical indices were proposed and the misclassification matrix was presented by rows or columns as producer's or consumer's accuracy respectively. Without dealing with these simple measures in great detail, the composite of this effort was to demonstrate that the attributes of a remotely-sensed image were just as crucial as the geometric fidelity. Some of the earliest work on statistically defensible tests for attribute accuracy was conducted by the GIRAS project at the US Geological Survey [FIT 81]. Although the remote-sensing community still has a somewhat distinct identity, the whole field of geographic information profited from this demonstration.

1.3.2. *Topology and logical consistency*

The concerns of data quality do not boil down to various accuracy measures either. It took some time to figure out that much of the effort in map compilation and digital data conversion was expended on tedious details that did not fit into the framework of most accuracy assessments. Interestingly, the criteria applied to these properties are more strict than either positional or attribute accuracy. If one polygon does not close properly, it can spread its color all over the map. The software may refuse to accept data coverages that are not completely "clean".

Though some of the early proponents of a topological approach to cartographic data structures [PEU 75] adopted a theoretical reasoning, the technique can find its origins in some error-checking procedures used in the verification of the address coding guides for the US Census [COO 98]. The topological data structure also had an important use in dealing with raw digitizer input [CHR 87]. The properties checked were not related strictly to accuracy, but more like the internal consistency of the representation. In drawing up a standard for cartographic data quality in the early 1980s, the issue of topological integrity was highly crucial.

1.3.3. *Fitness for use*

Modern information management has multiplied the opportunity for information to move from one user to another, eventually escaping the bounds of intended use. Professor Gersmehl [GER 85; CHR 95] is unlikely to be the first environmental scientist whose maps have been misinterpreted. Gersmehl recounted a story in which his dot-map of histosols was transformed inappropriately by one cartographer, and then applied to locate an energy research facility. With the increasing use of computer databases, the risks have been increased, and the nature of the problem has shifted. The stability required for a fixed-threshold test has vanished; also the set of concerns has broadened from strictly positional accuracy.

In a number of fields, the approach to quality evolved into a definition based on "fitness for use". This involves a change in the responsibilities. While a fixed threshold places the producer in charge of the test, the evaluation, and the assessment, these responsibilities do not all necessarily belong to the same group of people. For instance, the same product might be quite usable by one set of clients, but wildly inadequate for others. Under the fitness for use approach, the producer does not make any judgment, but simply reveals the results of some tests. The potential user evaluates these results according to the specific use intended.

Gersmehl's call for responsibility [GER 85] sounds crucial, but it is based on assumptions built into cartographic media. A mapmaker does bear much of the responsibility, as long as the user is considered a passive "bystander". However, the nature of database use, in its modern form, alters the simple assignment of responsibility. Modern database management offers a more flexible range of tools, unrestricted by the old pen and ink conventions of cartographic display (see [GOO 88; GOO 00] for a treatment of these changes). The illusion of comprehensive, consistent coverage, while perhaps necessary for map publication, has been replaced by more openness about limitations in the data. The database user must now bear as much responsibility as the originator because the user invokes many of the processing decisions and imparts much of the meaning. In an era of digital data, the original producer cannot foresee all potential users. Thus, it has become crucial for a data producer to record important aspects about the data so that users may make informed judgments regarding fitness for use.

The Spatial Data Transfer Standard (SDTS) [NIS 92] provides for a data-quality report containing five parts. The lineage section describes the source materials and all transformations performed to produce the final product. The other four sections describe tests that deal with various forms of accuracy. The positional accuracy test employs the established methods to describe the differences in coordinate measurements for "well-defined points" [ASP 89]. Attribute accuracy measures the fidelity of non-positional data, particularly classifications [CON 91]. Logical consistency refers to the internal relationships expected within the database. It can be checked without reference to the "real world". Completeness concerns the exhaustiveness of a collection of objects.

These five components can be repackaged in a number of more detailed categories, or regrouped into larger ones. The International Cartographic Association Commission on Data Quality [GUP 95] followed this five-part framework with a few elaborations [CHR 98]. The recent ISO TC211 standards build on the basis of SDTS, with the addition of explicit recognition of temporal accuracy, a topic considered in SDTS, but not given distinct status. While SDTS had a conceptual influence, the metadata standards promulgated by the US Federal Geographic Data Committee (FGDC) [FGD 94] have had a huge practical influence. The movement to create clearinghouses for "geo-spatial" information was started at a crucial moment just prior to the advent of the World Wide Web. Though it initially adopted a keyword style of searching more typical of library science, the collections of metadata have become a major resource in a networked world. The FGDC template provides many items that fulfill the intent of the five categories of data quality, some quite specific, others still as broad and unstructured as in the SDTS open text fields. To a large extent, the FGDC standard of 1994 simply implements the conceptual standard of 1984. In turn, the ISO metadata standard (and the Open GIS Consortium implementation) rearrange the items in some ways, but retain the same conceptual structure.

Despite attempts to harmonize the diversity of standards, the fitness for use perspective does not fit into the more common industrial standards of the modern world, particularly with ISO 9000 and its related standards, the darlings of the European Commission. ISO 9000 is written from the producer's perspective for industrial applications, and deals with documentation of procedure with little treatment of diversity in user interpretation. A motto of ISO 9000 is "deliver the expected quality, no more, no less". Yet, in the circumstances of a shared database of geographic information, one item may suit some users while not serving others. The whole ISO 9000 model is difficult to reconcile with fitness for use.

1.4. Elements of novelty

Is there anything new in the evaluation of data quality? Of course there is. The grand visions of a unified spatial data infrastructure emerging from the separate efforts of decentralized and dispersed actors have not proven automatic or easy. Goodchild [GOO 00] evokes the biblical story of the Tower of Babel, particularly in the entropic effects of no shared common language. Perhaps we are in an era where the limitations of the technological vision have become more evident. Perhaps some researchers, including Goodchild [GOO 04], Sui [SUI 04], Miller [MIL 04] and others in the GIScience movement hold a kind of nostalgia for a logical-empiricist science that discovers laws from the raw material of observations. The central problem is now seen as an ontological one (see [SMI 98] as an early example). Certainly the problem of data-quality assessment is rendered much more complicated if there is no common agreement on the objects of study. The creation of a Tower of Babel in the form of so many competing standards is perhaps evidence in support of the entropic post-apocalyptic message.

While some of the research community have tied themselves in knots over this argument about laws, ontology, and the trappings of science, there are some practical solutions that have received much less attention. I want to call attention to the Francophone concept of *"terrain nominal"* as a pragmatic recognition of multiple ontologies [SAL 95; DAV 97] – one that offers some useful solutions in place of the argumentation of the science wars.

The communication theories cited by Sui and Goodchild contend that the message is passed over a simple channel and adopt the cybernetic vision of Shannon and Weaver [SHA 64]. This emphasis on the instant of passing a message leaves out a large portion of the entire system that makes communication mean anything. It is too simple to call the surroundings "context" and leave that issue for another day. There is a specific set of agreements and understandings about practice that makes the messages work. The concept of *"terrain nominal"* encapsulates how geographic information is embedded inside agreements on how to look at the world. The word "nominal" evokes the nominalist philosophy that was the main opposition to overarching idealism during the debates of the scholars in the Middle Ages. Under the guise of "realism", some philosophers asserted the reality of things that could not be seen or measured. The more empiricist vein at the time was nominalist – asserting that grand words were just words. Of course philosophy and science have moved on to new issues over the past 700 years. Yet, there is still a division between those who tie science to some grand overarching principles and those who profess more modest goals. Kuhn [KUH 70] made the observation that science was not a smooth path towards refined knowledge, but was punctuated by revolutions. These revolutions involve substituting one "paradigm" with another, rendering the measurements and theories incommensurable. The concept of *"terrain nominal"* fits into this anti-realist approach to scientific measurement.

The central concept of *"terrain nominal"* is that a test for accuracy cannot be performed in some naïve way. You cannot simply go to the field and return with a measurement of greater accuracy. The ultimate truth does not come from an external source, but must remain inside the same set of meanings as the measurement to be tested. Thus, you do not visit the terrain without the filters of an established approach to classifying objects, describing their relationships, and measuring their properties. The test is not performed in the "real" world (whatever that might mean), but in a "nominal" world where objects are defined according to a specification and where techniques of measurement are defined. In the Anglophone world, this pragmatic proceduralist term has been translated as "abstract universe", "perceived reality", and other terms that give an exactly opposite philosophical impression of a realist higher world of abstract forms that are independent of the observer. The *"terrain nominal"* is anti-essentialist, recognizing that confronted with the variety of the world, there have to be specifications that will lead to reproducible results.

My discussion of this term has been phrased in the terminology of philosophy, a field that is rarely invoked in the practical engineering of accuracy assessment. Yet, disputes about data-quality assessment can be traced to well-established philosophical positions. Rather than reinventing the whole discourse, it makes some sense to understand the quality assessment of geographic information inside a larger framework. The first result of this process is to find, as Robinson and Petchenik [ROB 76] commented, that maps are glibly used as metaphors when philosophers talk about science, models, and theories. This reflexivity does not have to lead to circularity [SIS 01]. Maps and geographic information are indeed, in their variety and complexity, good examples to use to describe the complexity of scientific representations. Some maps do attempt to be passive mirrors of the world, but not in the universal sense expected. As Denis Wood [WOO 92] points out, maps reflect power relations between people and organizations. They often reflect not only what is to be seen, but who is doing the seeing and who is paying for their work. As Hutchin's study of navigation as "cognition in the wild" [HUT 95] demonstrates, the accuracy of a nautical chart should not be examined outside the framework of the navigation system in which the chart is used. Positions on the chart are designed for angular measurements, not distance relationships. The scale on a Mercator projection varies continuously, while the angles are perfectly preserved. Similarly, any assessment of the accuracy of the famous map of the London Underground must reflect the topological principles that are preserved in that most useful, yet most geometrically distorted, representation.

Where does this discussion lead us? Using the starting point of the humble term "terrain nominal", I want to demonstrate that accuracy assessment requires some humility about the source of "truth". This is not to say that there is no concept of reality and truth, just that each term has to be carefully defined and contained inside a system of meaning. An absolute truth, a universal test, is demonstrably unattainable. Yet, all is not lost in some relativist miasma. The fact that tests are relative to a specific set of specifications has been long recognized as a simple necessity in training personnel. Manuals of procedure are treated as boring, but they actually provide the key to build the ontologies that have become the rage of applied computer science. The most humble elements of our practices may turn out to have philosophical implications.

In conclusion, this chapter began with a story of the development of the study of data quality. This once-neglected field has become a central concern in the far-flung practice of geographic information technology. The development of this field has sparked considerable debate, and I selected one concept, "terrain nominal" as an indication of the philosophical turn that can be anticipated.

1.5. References

[ASP 89] AMERICAN SOCIETY FOR PHOTOGRAMMETRY AND REMOTE SENSING, "Interim Standards for large scale line maps", *Photogrammetric Engineering and Remote Sensing*, vol. 55, p. 1038-1040, 1989.

[BUR 47] BUREAU OF THE BUDGET, *National Map Accuracy Standards*, GPO, Washington DC, 1947.

[CHR 84] CHRISMAN N.R., "The role of quality information in the long-term functioning of a geographic information system", *Cartographica*, vol. 21, p. 79-87, 1984.

[CHR 87] CHRISMAN N.R., "Efficient digitizing: advances in software and hardware", *International Journal of Geographical Information Systems*, vol. 1, p. 265-277, 1987.

[CHR 95] CHRISMAN N.R., "Living with error in geographic data: truth and responsibility", *Proceedings GIS'95*, p. 12-17, GIS World, Vancouver, 1995.

[CHR 98] CHRISMAN N.R., "Review of elements of data quality", *Cartography and Geographic Information Systems*, vol. 25, p. 259-260, 1998.

[CLO 02] CLOUD J., "American cartographic transformations during the Cold War", *Cartography and Geographic Information Science*, vol. 29, p. 261-282, 2002.

[CON 91] CONGALTON R.G., "A review of assessing the accuracy of classifications of remotely sensed data", *Remote Sensing of Environment*, vol. 37, p. 35-46, 1991.

[COO 98] COOKE D.F., "Topology and TIGER: the Census Bureau's contribution", in Foresman T.W. (ed.), *The History of Geographic Information Systems: Perspectives from the Pioneers*, p. 57-58, Upper Saddle River NJ, Prentice Hall, 1998.

[DAV 97] DAVID B., FASQUEL P., "Qualité d'une base de données géographique: concepts et terminologie", *Bulletin d'information de l'Institut Géographique National*, no. 67, IGN, France, 1997.

[FIT 81] FITZPATRICK-LINS K., "Comparison of sampling procedures and data analysis for a land use and land cover map", *Photogrammetric Engineering and Remote Sensing*, vol. 47, p. 343-351, 1981.

[FGD 94] FGDC, *Content Standards for Digital Geospatial Metadata*, Reston VA, Federal Geographic Data Committee, 1994.

[FGD 98] FGDC, *National Standard for Spatial Data Accuracy*, FGDC-STD 007.3 1998, Reston VA, Federal Geographic Data Committee, 1998.

[GER 85] GERSMEHL P.J., "The data, the reader, and the innocent bystander – a parable for map users", *Professional Geographer*, vol. 37, p. 329-334, 1985.

[GOO 88] GOODCHILD M.F., "Stepping over the line: technological constraints and the new cartography", *The American Cartographer*, vol. 15, p. 311-320, 1988.

[GOO 00] GOODCHILD M.F., "Communicating geographic information in a digital age", *Annals of the Association of American Geographers*, vol. 90, p. 344-355, 2000.

[GOO 04] GOODCHILD M.F., "The validity and usefulness of laws in geographic information science and geography", *Annals of the Association of American Geographers,* vol. 94, p. 300-303, 2004.

[GUP 95] GUPTILL S.K., MORRISON J. (eds.), *Elements of Spatial Data Quality,* Oxford, Elsevier, 1995.

[HAY 77] HAYES G.E., ROMIG H.G., *Modern Quality Control,* Encino CA, Bruce, 1977.

[HUT 95] HUTCHINS E., *Cognition in the Wild,* Cambridge MA, MIT Press, 1995.

[KUH 70] KUHN T.S., *The Structure of Scientific Revolutions,* Chicago, University of Chicago Press, 1970.

[MAC 90] MACKENZIE D.A., *Inventing Accuracy: An Historical Sociology of Nuclear Missile Guidance,* Cambridge MA, MIT Press, 1990.

[MIL 04] MILLER H.J., "Tobler's first law and spatial analysis", *Annals of the Association of American Geographers,* vol. 94, p. 284-289, 2004.

[NIS 92] NATIONAL INSTITUTE OF STANDARDS AND TECHNOLOGY, *Spatial Data Transfer Standard,* Washington DC, National Institute of Standards and Technology, Department of Commerce, 1992.

[PEU 75] PEUCKER T.K., CHRISMAN N.R., "Cartographic data structures", *The American Cartographer,* vol. 2, p. 55-69, 1975.

[ROB 76] ROBINSON A.H., PETCHENIK B.B., *The Nature of Maps: Essays Toward Understanding Maps and Mapping,* Chicago, University of Chicago Press, 1976.

[SAL 95] SALGÉ F., "Semantic accuracy", in Guptill S.K., Morrison J. (eds.), *Elements of Spatial Data Quality,* Oxford, Elsevier, 1995.

[SHA 64] SHANNON C.E., WEAVER W., *The Mathematical Theory of Communication,* University of Illinois, Urbana, Illinois, USA, 1964.

[SIS 01] SISMONDO S., CHRISMAN N.R., "Deflationary metaphysics and the natures of maps", *Philosophy of Science,* vol. 68, p. S38-S49, 2001.

[SMI 98] SMITH B., MARK D.M., "Ontology and geographic kinds", *Proceedings of Spatial Data Handling 98,* p. 308-320, IGU, Vancouver BC, 1998.

[SUI 04] SUI D.Z., "Tobler's first law of geography: a big idea for a small world?", *Annals of the Association of American Geographers,* vol. 94, p. 269-277, 2004.

[WOO 92] WOOD D., *Power of Maps,* New York, Guilford, 1992.

Chapter 2

Spatial Data Quality: Concepts

2.1. Introduction

"All models are wrong, but some are useful" [BOX 76].

Geospatial data are a "model of reality" [LON 01], a reasoned and simplified representation of this complex reality. Every map or database is therefore a model, produced for a certain purpose, in which certain elements deemed non-essential have been simplified, grouped, or eliminated, in order to make the representation more understandable and thereby encourage the process of information communication. A set of factors is added to the modeling process that may be the source of errors in the data produced, related to the technology (for example, precision of measuring devices, algorithms used to manipulate data) and to humans involved in the creation of data (for example, visual identification of objects in images).

All geospatial data are then, at different levels, imprecise, inaccurate, out-of-date, incomplete, etc.[1] These divergences between reality and representation may however be deemed acceptable within certain applications. Figure 2.1 presents the same phenomenon from the real world (that is, roads) represented in different municipal and governmental geospatial databases (that is, cadastral and topographical maps at scales ranging from 1:1,000 to 1:250,000). None of these representations corresponds strictly to reality, but these models represent the same reality at different levels of abstraction, to meet different needs.

Chapter written by Rodolphe DEVILLERS and Robert JEANSOULIN.

1 With the exception of certain specific cases where data are legally recognized as reality (for example, legal cadastre of certain countries).

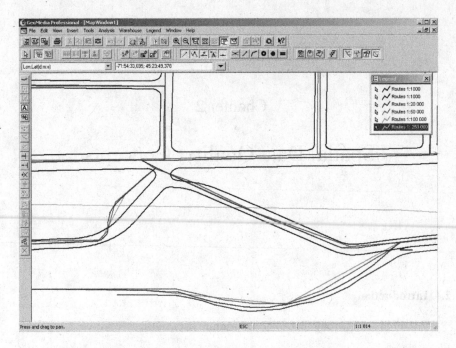

Figure 2.1: *Overlaying of road network from datasets at different scales*

A perfect representation of the territory would probably require a map at 1:1 scale, as discussed with humor by Lewis Carroll in "Sylvie and Bruno" [CAR 82]:

> "That's another thing we've learned from your Nation," said Mein Herr, "map-making. But we've carried it much further than you. What do you consider the largest map that would be really useful?"

> "About six inches to the mile."

> "'Only six inches!'" exclaimed Mein Herr. "We very soon got to six yards to the mile. Then we tried a hundred yards to the mile. And then came the grandest idea of all! We actually made a map of the country, on the scale of a mile to the mile!"

> "Have you used it much?" I enquired.

> "It has never been spread out, yet," said Mein Herr, "the farmers objected: they said it would cover the whole country, and shut out the sunlight! So we now use the country itself, as its own map, and I assure you it does nearly as well."

The famous expression "the map is not the territory" that first appeared in a paper that Alfred Korzybski gave in New Orleans in 1931 [KOR 31] also illustrates the process of abstraction occurring when creating any model. In spite of many maps and many models produced before and after August 2005, the inhabitants of the city of New Orleans are now totally convinced that the map is not their territory!

All geospatial data are named and designed even before being acquired, even with the help of an instrument that is apparently neutral. Bachelard spoke of instruments as "materialized theories" and a number of philosophers speak of facts today as "theory-loaded" [HAN 58].

Regardless of whether it is called models, data, facts, instruments, or observations, a filter exists causing a deformation between the information supplied and the portion of reality that it is supposed to represent. Geospatial data do not escape this filter, which deforms the data in several ways:

– the result can differ to a certain degree from what the theory was expecting (error, imprecision, incompleteness);

– theory cannot reflect exactly what its designer had intended (ambiguity);

– the theory on which the observation is based is not necessarily in agreement with the intentions of the user (discord).

The problem of data quality is therefore situated at the heart of geomatics, omnipresent in all applications, and has a major impact on the decisions based on these data.

Over the last 20 years, the following two revolutions have increased interest in the field of spatial data quality. The first is the transition from paper to digital maps. While users seem more aware of problems with the quality of paper maps, the digital format gives users an appearance of great accuracy since they can measure distances very precisely or visualize data at a very high level of detail. The second revolution is the development of the Internet and other forms of communication, facilitating the exchange of data between individuals and organizations. We have passed from one paradigm where we often had a user (or an organization) that would produce data for its own usage, to a situation where multiple users can download data of variable quality to their terminal, and, often unknowingly, integrate them into their application.

This chapter attempts to present certain general concepts related to spatial data quality. In section 2.2 we present different sources and types of errors that can be at the source of quality problems. Then we will present, in section 2.3, a brief history of the field of data quality and different definitions of the concept of quality with an emphasis on internal and external quality.

2.2. Sources and types of errors

Heuvelink [HEU 99] defines error as the difference that exists between a reality and the representation of this reality. Errors will directly influence the internal quality of the data produced (see section 2.3 for a description of the concept of internal quality). Different authors have proposed classifications for these errors and their sources. Beard [BEA 89] proposed to classify errors into three categories: errors related to the acquisition and the compilation of data (*source error*), those related to processing (*process error*), and those related to the use of data (*use error*). Hunter and Beard [HUN 92] extend this classification of errors, adding the type of errors to the sources (see Figure 2.2). In this model, the three categories of source errors create primary errors (positional errors and attribute errors) and secondary errors (logical consistency and completeness). All of these errors form the global error attached to the final product.

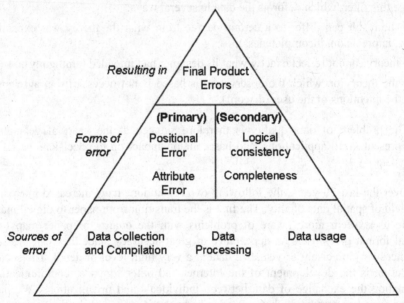

Figure 2.2: *Classification of errors in GIS [HUN 92]*

More specifically, Collins and Smith [COL 94], for example, separated the errors according to different phases, extending from data collection to data use:

– *Data collection*: inaccuracies in field measurements; inaccurate equipment; incorrect recording procedures; errors in analysis of remotely sensed data.

– *Data input*: digitizing error; nature of fuzzy natural boundaries; other forms of data entry.

– *Data storage*: numerical precision; spatial precision (in raster systems).

– *Data manipulation*: wrong class intervals; boundary errors; spurious polygons and error propagation with overlay operations.

– *Data output*: scaling; inaccurate output device.

– *Data usage*: incorrect understanding of information; incorrect use of data.

2.3. Definitions of the concept of quality

The term "quality" comes from the Latin "*qualitas*", inspired from the Greek "ποιον" (one of the main "categories" of Aristotle) and attributed to the philosopher Cicero, based on "*qualis*" meaning "what" (that is, the nature of something). Quality can be paraphrased as "what is it?", or as the answer to the question that immediately follows the first ontological question "is there?". Is there something else and what is that something? At several places in this book, we will examine how questions about quality are, above all, questions about "the quality of what" (see the chapters on external quality).

"Quality", as a category, deals with unquantifiable characteristics, as opposed to "quantity". This meaning of quality is found, for example, in the expression "to act in the quality of" (see Appendix), as well as in philosophy where quality can be defined as the "aspect of experience that differs specifically from any other aspect and, in that, allows you to distinguish this experience" [OLF 04].

From the beginning of the 20th century, the term "quality" was mainly used in the field of production of goods, to enable the creation of quality products ("quality" meaning *the sense of excellence*). The engineer F.W. Taylor [TAY 11], to whom is attributed the first works related to the field of quality, described a set of principles, dealing with work management, to improve the quality of goods produced. These principles, which form the basis of "Taylorism", are, according to their author, the basis for a new science of work. Ford successfully applied these principles in its automobile assembly plants ("Fordism"). Later, at the end of World War II, the concept of total quality management (TQM) appeared, based on the work of Deming and other important players in the field of quality (for example, Juran, Feigenbaum, and later, Shingo, Taguchi, and Crosby). These works, which were very different from those by Taylor, were first applied in Japan, and encouraged the return of industrial growth in the country. The concepts related to quality management were then standardized by ISO (Technical Committee 176 – 1979) and resulted in ISO 9000 in 1987.

If all these authors agree on the importance of quality, their definitions of what quality is vary greatly and no consensus exists in the community on a single definition of quality, although certain definitions, such as the those used by ISO, are generally accepted. Therefore, for some people, a quality product is exempt from error, or is a product that is in conformity with the specifications used. For others it is a product that meets consumer expectations.

Several authors grouped these definitions into two broad groups: *internal quality* (products that are exempt from error) and *external quality* (products that meet user needs). These two definitions of quality are presented in sections 2.3.1 and 2.3.2.

In the field of geomatics, these two groups of definitions of quality are also used. If, when we speak of spatial data quality, most people think mainly about spatial data accuracy (a criterion included in internal data quality), the more official definitions of quality correspond, however, to external quality. In fact, this duality is often a source of confusion in the literature.

Figure 2.3 presents in a simplified way the concepts of internal and external quality. On one side, internal quality corresponds to the level of similarity that exists between "perfect" data to be produced (what is called "nominal ground") and the data actually produced. On the other side, external quality corresponds to the similarity between the data produced and user needs.

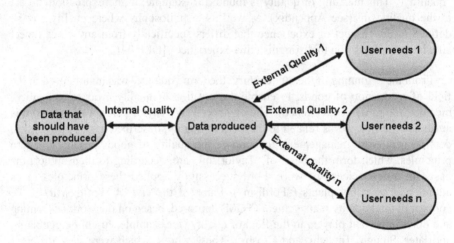

Figure 2.3: *Concepts of internal and external data quality*

The concept of quality implies that the data produced are not perfect and that they differ from data that should have been produced. Therefore, different types of errors can appear during the data production process (see section 2.2 on errors). Based on a classic humorous illustration from the field of information technology,

Figure 2.4 presents the concepts of internal and external quality at different stages of data/system production.

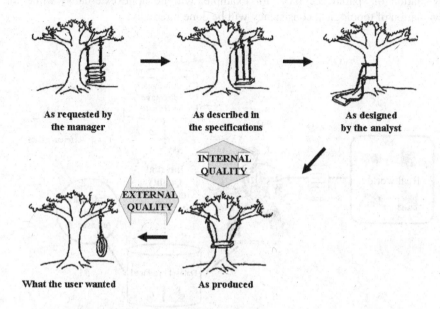

Figure 2.4: *Internal and external quality in a production process*

2.3.1. *Internal quality*

Internal quality corresponds to the level of similarity that exists between the data produced and the "perfect" data that should have been produced (that is, data produced without error) (see Figure 2.5). These perfect data are often called "nominal ground" [DAV 97] or "universe of discourse" [ISO 02]. Nominal ground is defined by David and Fasquel [DAV 97] as "an image of the universe, at a given date, through the filter defined by the specifications of the product". The specifications of a product are defined as "a document that prescribes the requirements to which the product must conform" [DAV 97]. The specifications are a set of rules and requirements that define the way of passing from the real world to the data. The specifications include, for example, the definitions of objects to be represented, the geometries to be used to represent each type of object (for example, point, line, polygon), the attributes that describe them, and the possible values for these attributes. In practice, the evaluation of internal quality does not use the nominal ground (see Figure 2.5) that has no real physical existence since it is an "ideal" dataset, but uses a dataset of greater accuracy than the data produced, which is called "control data" or "reference data" (see Chapter 11). The evaluation of internal quality includes an external part (comparison with reference data), as well as an internal part, depending on the elements of quality to be verified. The evaluation of spatial accuracy, for example, will be done externally, while the evaluation of topological consistency will be done internally.

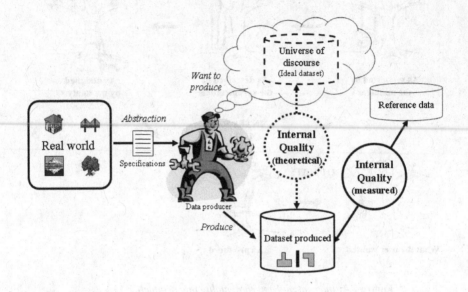

Figure 2.5: *Concept of internal quality*

Internal quality can be described using different criteria. Overall, there has been a consensus about the criteria for nearly 20 years and the criteria have been used by the main standards in geomatics (for example, ISO, FGDC, CEN). The ISO 19113 standard [ISO 02] on geospatial data quality recommends the use of the following criteria:

   – *Completeness*: presence and absence of features, their attributes and relationships.

   – *Logical consistency*: degree of adherence to logical rules of data structure, attribution, and relationships (data structure can be conceptual, logical or physical).

   – *Positional accuracy*: accuracy of the position of features.

   – *Temporal accuracy*: accuracy of the temporal attributes and temporal relationships of features.

   – *Thematic accuracy*: accuracy of quantitative attributes and the correctness of non-quantitative attributes and of the classifications of features and their relationships.

These criteria correspond to "elements" of quality, each of these elements being composed of sub-elements (see Chapter 10 for more details).

The internal quality of data can be improved in different ways during the data creation process (see Chapters 8 and 9) and can be evaluated *a posteriori* (see Chapters 10 and 11) to eventually be communicated to users (see Chapters 10 and 12).

2.3.2. *External quality*

The concept of external quality corresponds to the level of concordance that exists between a product and user needs, or expectations, in a given context. The concept of external quality can therefore be associated with the adjective "better" as it is commonly used. If several people were asked to name the best brand of car, the best book, or the best cake, there would be a wide range of different answers. Each person's choices would depend on variable criteria, and be based on their preferences, past experiences, etc. Therefore, if you wanted to compare two cars, a Ford and a Rolls-Royce for example, to find out which car had better internal quality, the Rolls-Royce would probably be chosen. However, with regard to external quality, the Rolls-Royce could be perceived as too expensive to buy and to insure, costly to repair, attractive to car thieves, etc. Therefore, from the average consumer's point of view, a Ford would probably be preferred (and would have better external quality). In the same way, a set of data of great internal quality representing the hydrology of a region, created by following the best quality control, could be very useful (that is, good external quality) to an environmental expert and almost useless (that is, poor external quality) to a land surveyor.

Therefore, the concept of external quality implies that quality is not absolute and the same product can be of different quality to different users (see Figure 2.3).

The concept of external quality is generally recognized as the definition of quality in the largest sense. ISO [ISO 94] defines quality as the "totality of characteristics of a product that bear on its ability to satisfy stated and implied needs". External quality is often defined as "fitness for use" or "fitness for purpose". The definition of "fitness for use", proposed by Juran [JUR 74], has appeared in geographic information in the American standard NCDCDS (1982), as well as in an article by Chrisman [CHR 83].

The evaluation of external quality can imply criteria that describe internal quality. To evaluate whether a dataset meets our needs, we can check to see if the data represent the territory required at an appropriate date include the necessary objects and attributes, but also, if the data have sufficient spatial accuracy or completeness, etc. These last two criteria describe internal quality.

A few authors have suggested criteria that can be used to evaluate external quality. Among these authors, Wang and Strong [WAN 96] identified four dimensions for external quality, based on a survey conducted of approximately 350 users of non-geospatial data:

– *Intrinsic Data Quality*: believability; accuracy; objectivity; reputation.

– *Contextual Data Quality*: value-added; relevancy; timeliness; completeness; appropriate amount of data.

– *Representational Data Quality*: interpretability; ease of understanding; representational consistency; concise representation.

– *Accessibility Data Quality*: accessibility; access security.

For geospatial data, Bédard and Vallière [BED 95] proposed six characteristics to define the external quality of a geospatial dataset:

– *Definition*: to evaluate whether the exact nature of a data and the object that it describes, that is, the "what", corresponds to user needs (semantic, spatial and temporal definitions).

– *Coverage*: to evaluate whether the territory and the period for which the data exists, that is, the "where" and the "when", meet user needs.

– *Lineage*: to find out where data come from, their acquisition objectives, the methods used to obtain them, that is, the "how" and the "why", and to see whether the data meet user needs.

– *Precision*: to evaluate what data is worth and whether it is acceptable for an expressed need (semantic, temporal, and spatial precision of the object and its attributes).

– *Legitimacy*: to evaluate the official recognition and the legal scope of data and whether they meet the needs of *de facto* standards, respect recognized standards, have legal or administrative recognition by an official body, or legal guarantee by a supplier, etc.;

– *Accessibility*: to evaluate the ease with which the user can obtain the data analyzed (cost, time frame, format, confidentiality, respect of recognized standards, copyright, etc.).

While methods exist to evaluate internal data quality (see Chapter 11), the evaluation of external quality remains a field that has been explored very little (see Chapters 13 and 14).

2.4. Conclusion

Quality is of utmost concern when examining geospatial data, whether in the process of geographical data production or in the context of a real-life usage. The term "quality" is defined and used in different and sometimes contradictory ways by the geospatial data community. Chapter 3 clarifies the diversity of the terms that are related to questions of quality.

This chapter was intended to present certain basic concepts about quality, concentrating on the distinction that is made between "internal quality" and "external quality". These two very different visions of quality reflect the two perspectives that are usually found in literature about quality. The remaining chapters in this book will explain these terms and illustrate them in different contexts.

2.5. References

[BEA 89] BEARD M.K., "Use error: the neglected error component", *Proceedings of AUTO-CARTO 9*, 2–7 April 1989, Baltimore, USA, p 808–17.

[BED 95] BEDARD Y. and VALLIERE D., *Qualité des données à référence spatiale dans un contexte gouvernemental*, technical report, 1995, Quebec City, Laval University.

[BOX 76] BOX G.E.P., "Science and statistics", *Journal of the American Statistical Association*, vol 71, 1976, p 791–99.

[CAR 82] CARROLL L., *The Complete, Fully Illustrated Works*, Gramercy Books, 1982.

[CHR 83] CHRISMAN N.R., "The role of quality in the long-term functioning of a geographic information system", *Proceedings of AUTO-CARTO 6*, vol 2, October 1983, Ottawa, p 303–21.

[COL 94] COLLINS F.C. and SMITH J.L., "Taxonomy for error in GIS", *Proceedings of the International Symposium on the Spatial Accuracy of Natural Resource Data Bases*, 16–20 May 1994, Williamsburg, USA, ASPRS, p 1–7.

[DAV 97] DAVID B. and FASQUEL P., "Qualité d'une base de données géographique: concepts et terminologie", *Bulletin d'information de l'IGN*, number 67, 1997, IGN, France.

[HAN 58] HANSON N.R., *Patterns of Discovery*, 1958, Cambridge, Cambridge University Press.

[HEU 99] HEUVELINK G.B.M., "Propagation of error in spatial modelling with GIS", in *Geographical Information Systems: Principles and Technical Issues*, 1999, vol 1, 2nd edn, Longley P.A., Goodchild M.F., Maguire D.J. and Rhind D.W. (eds), New York, John Wiley & Sons, p 207–17.

[HUN 92] HUNTER G.J. and BEARD M.K., "Understanding error in spatial databases", *Australian Surveyor*, 1992, vol 37, number 2, p 108–19.

[ISO 94] ISO 8402, *Quality Management and Quality Assurance – Vocabulary*, International Organization for Standardization (ISO), 1994.

[ISO 02] ISO/TC 211, 19113 *Geographic Information – Quality Principles*, International Organization for Standardization (ISO), 2002.

[JUR 74] JURAN J.M., GRYNA F.M.J. and BINGHAM R.S., *Quality Control Handbook*, 1974, New York, McGraw-Hill.

[KOR 31] KORZYBSKI A., "A Non-Aristotelian System and its Necessity for Rigour in Mathematics and Physics", Meeting of the American Association for the Advancement of Science, American Mathematical Society, New Orleans, Louisiana, 28 December 1931, reprinted in *Science and Sanity*, 1933, pp 747–61.

[LON 01] LONGLEY P.A., GOODCHILD M.F., MAGUIRE D.J. and RHIND D.W. (eds), *Geographical Information Systems and Science*, 2001, New York, John Wiley and Sons.

[OLF 04] OFFICE QUÉBÉCOIS DE LA LANGUE FRANÇAISE, 2004, http://www.olf.gouv.qc.ca.

[TAY 11] TAYLOR F.W., *The Principles of Scientific Management*, 1911, New York, Harper and Row.

[WAN 96] WANG R.Y. and STRONG D.M., "Beyond accuracy: what data quality means to data consumers", *Journal of Management Information Systems*, 1996, vol 12, p 5–34.

Chapter 3

Approaches to Uncertainty in Spatial Data

3.1. Introduction

Geographical information systems (GIS) are designed to handle large amounts of information about the natural and built environments. Any such large collection of observational information is prone to uncertainty in a number of forms. If that uncertainty is ignored there may be anything from slightly incorrect predictions or advice, to analyses that are completely logical, but fatally flawed. In either case, future trust in the work of the system or the operator can be undermined. It is therefore of crucial importance to all users of GIS that awareness of uncertainty should be as widespread as possible. Fundamental to that understanding is the nature of the uncertainty – the subject of this chapter.

This chapter explores the conceptual understanding (modeling) of different types of uncertainty within spatial information (Figure 3.1). At the heart of the issue of uncertainty is the problem of defining the class of object to be examined (for example, soils), the individual object to be observed (for example, soil map unit), the property or properties to be measured (for example, field description), and the processing of those observations (for example, numerical taxonomy). These stages can be prone to problems of definition, misunderstanding, doubt, and error all of which can contribute in various ways to the uncertainty in the information. Above all, the failure to communicate and understand precisely the decisions that are made can lead to, at worst, the misuse of information or the wrong decision. Once the

Chapter written by Peter FISHER, Alexis COMBER and Richard WADSWORTH.

conceptual modeling identifies whether the class of objects to be described is *well* or *poorly* defined, the nature of the uncertainty follows:

– If both the class of object and the individual object are *well defined*, and the observation is objective to make, then the uncertainty is caused by errors and is probabilistic in nature.

– If the class of object or the individual object is *poorly defined*, then additional types of uncertainty may be recognized. Recent developments in research mean that all areas have been explored by GIS researchers:

  - If the uncertainty is due to the *poor* definition of the class of object or individual object, then definition of a class or set within the universe is a matter of vagueness, which can be treated by fuzzy set theory, although other formalisms, such as super-valuation, are also suggested.

  - Uncertainty may also arise because of ambiguity (the confusion over the definition of sets within the universe) owing, typically, to differing classification systems. This also takes two forms [KLI 95], namely:

    - where one object or individual is clearly defined, but is shown to be a member of two or more different classes under differing schemes or interpretations of the evidence, then discord arises which can be related to the classification scheme for the information. Ultimately such discord is treated as a matter of the semantics or ontology of the database, and solutions to the problem rely on an understanding of the semantics of the classification schemes and can be formalized through artificial intelligence (AI) methods, including Dempster Schafer's Theory of the Evidence (Comber *et al.*, Chapter 7);

    - where the process of assigning an object to a class is open to interpretation, then the problem is non-specificity, which, like discord, can be treated by a number of AI methods, including endorsement theory (Comber *et al.*, Chapter 7), as well as fuzzy-set theory.

This list of uncertainty concepts is probably not exhaustive, but most if not all the current research into uncertainty can be associated with one of other concept, or with the possibly vague boundaries between them.

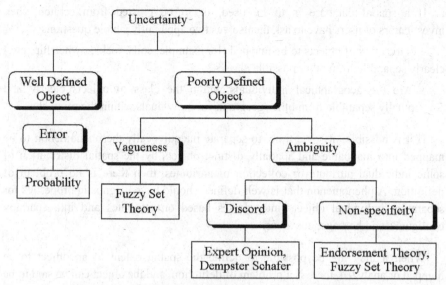

Figure 3.1. *A conceptual model of uncertainty in spatial data*
(after [FIS 99], adapted from [KLI 95]: 268, with revisions)

3.2. The problem of definition

The principal issue of geographical uncertainty is the possible mismatch of understanding of what is to be, or has been measured, as between the data collector and any users. Within geographical information, that uncertainty roams over the three principle dimensions of description, namely uncertainty in measurement of attributes, of space, and of time. In order to define the nature of the uncertainty of an object within the dimensions of space and time, a decision must be made as to whether or not it is clearly and meaningfully separable from other objects in whichever dimension is of interest – ideally it will be separable in both space and time. This is a complex intellectual process, and one that draws on the history and the critical appraisal of subject-specific scientists. This conceptual model has been complicated and muddied by current paradigms that influence the perception of geographical information. Foremost among those is the historical necessity of simplification of information for map production – what Fisher [FIS 98b] calls the paradigm of "production cartography". Equally important, however, are the concepts of classification, commonly based on hierarchies, in which objects must fall into one class or another, and of computer database models where instances of objects are treated as unique individuals which can be the basis of a transaction and which are often organized hierarchically.

If a spatial database is to be used, or to be created from scratch, then investigators or users have to ask themselves two apparently simple questions:

– Is the class of objects to be mapped (for example, soils, rocks, ownership, etc.) clearly separable from other possible classes?

– Are the geographical individuals within the class of objects clearly and conceptually separable from other geographical individuals within the same class?

If it is possible, at some time, to separate unequivocally the phenomenon to be mapped into mappable and spatially distinct objects by the spatial distribution of some individual attribute or collection of attributes, then there is no problem of definition. A phenomenon that is well defined should have diagnostic properties for separating individual objects into classes based on attributes and into spatially contiguous and homogenous areas.

When it may not be possible to define the spatial extent of an object to be mapped or analyzed, there is a problem of definition, and the object can be said to be "vague" [WIL 94]. In this circumstance, while specific properties may be measured and these measurements may be very precise, no combination of properties allows the unequivocal allocation of individual objects to a class, or even the definition of the precise spatial extent of the objects. Most spatial phenomena in the natural environment share this problem of definition, at least to some extent. Error analysis on its own does not help with the description of these classes, although any properties, which are measured, may be subject to errors just as they are in other cases.

3.2.1. *Examples of well-defined geographical objects*

Census geographies tend to be well-defined *fiat* objects [SMI 01]. They consist of a set of regions, each with precise boundaries created by the census organization according to either their own needs or a set of electoral units (and usually both at different levels of hierarchy; [OPE 95]). The areas at the lowest level of enumeration (enumeration districts, city blocks, etc.) are grouped with specific instances of other areas at the same level to make up higher level areas which are themselves grouped with other specific areas to form a complete and rigid hierarchy [MAR 99]. The attributes to be counted within the areas are based on properties of individuals and households, but even the definition of household may vary between surveys [OFF 97].

A second example of a well-defined geographical phenomenon in Western societies is land ownership. The concept of private ownership of land is fundamental to these societies, and therefore the spatial and attribute interpretation of that concept is normally quite straightforward in its spatial expression. The boundary between land parcels is commonly marked on the ground, and shows an abrupt and total change in ownership.

Well-defined geographical objects are created by human beings to order the world that they occupy. They exist in well-organized and established political realms, as well as legal realms. Some other objects in our built and natural environments may seem to be well defined, but they tend to be based on a single measurement, and close examination frequently shows them to be less well defined. For example, the land surface seems well defined, but the consequence of increasingly precise measurement is upsetting this assumption. Most, if not all, other geographical phenomena are to some extent similarly poorly defined.

3.2.2. *Examples of poorly defined geographical objects*

Poorly-defined objects are the rule in natural resource mapping. The allocation of a patch of woodland to the class of oak woodland, as opposed to any other candidate vegetation type, is not necessarily easy. It may be that in that region a threshold percentage of trees need to be oak for the woodland to be considered 'oak', but what happens if there is 1% less than that threshold? Does it really mean anything to say that the woodland needs to belong to a different category? Indeed, the higher-level classification of an area as woodland at all has the same problems (Figure 3.2; [COM 04b; LUN 04]). Mapping the vegetation is also problematic since in areas of natural vegetation there are rarely sharp transitions from one vegetation type to another, rather an intergrade zone or ecotone occurs where the dominant vegetation type is in transition [MOR 93]. The ecotone may occupy large tracts of ground. The attribute and spatial assignments may follow rules, and may use indicator species to assist decisions, but strict deterministic rules may trivialize the classification process without generating any deeper meaning.

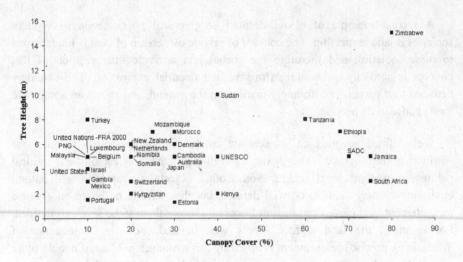

Figure 3.2 *The minimum physical requirements of a "forest"; data from [LUN 04], figure from [COM 04b]. Note: most countries do not actually use these variables in defining their forests, but this is the result of their definitions*

Within natural resource disciplines, the conceptualization of mappable phenomena and the spaces they occupy is rarely clear-cut, and is still more rarely achieved without invoking simplifying assumptions. In forestry, for example, tree stands are defined as being clearly separable and mappable, and yet trees vary within a stand by species density, height, etc. and often the spatial boundary between stands is not well defined [EDW 94]. Although theorists may recognize the existence of intergrades, the conceptual model of mapping used in this and other natural resource disciplines accepts the simplification into clear classes, and places little importance on the intergrades, although the significance has not been assessed. The interest in intergrades as boundaries is not a preserve of natural resource scientists, however, and in any discussion of urban and political geography considerable attention needs to be paid to these concepts [THU 00].

3.3. Error

If an object is conceptualized as being definable in both attribute and spatial dimensions, then it has a Boolean occurrence; any location is either part of the object, or it is not. Yet within GIS, for a number of reasons, the assignment of an object or location to the class may be expressed as a probability, or indeed the location itself can be expressed probabilistically. There are any number of reasons why this might be the case. Three are briefly discussed here.

Errors occur within any database, and for any number of reasons (see Table 3.1). The easiest errors to handle are those associated with measurement, because well-advanced error analysis procedures have been developed [HEU 99; TAY 82]. If a precise measurement of a property could be made, then by repeated measurements of that property of the object it would be possible to estimate the distribution of the error in its measurement, and thus develop a full model of the measurement error. This is the basis of the "root mean square error" reporting of error in a digital elevation model. On the other hand, there are many other instances in which such reductionist measures of error are over-simplistic and aspatial, and fail to identify the spatial distribution of the error in GIS-based modeling [FIS 98a].

Type of Error	Cause of error
Measurement	Measurement of a property is erroneous.
Assignment	The object is assigned to the wrong class because of measurement error by the scientist in either the field or laboratory or by the surveyor.
Class Generalization	Following observation in the field, and for reasons of simplicity, the object is grouped with objects possessing somewhat dissimilar properties.
Spatial Generalization	Generalization of the cartographic representation of the object before digitizing, including displacement, simplification, etc.
Entry	Data are miscoded during (electronic or manual) entry in a GIS.
Temporal	The object changes character between the time of data collection and the time of database use.
Processing	In the course of data transformations an error arises because of rounding or algorithm error.

Table 3.1. *Common reasons for a database being in error ([FIS 99])*

A further example of aspatial error description is the confusion matrix which shows the cover-type actually present at a location and cross-tabulated against the cover-type identified in the image classification process. Typically, the matrix is generated for the whole image. It reports errors in allocation of pixels to cover-types, [CON 83], but the precise interpretation of either the classification process or the ground information may not be clear cut and any possible variation in space is completely ignored.

Probability has been studied in mathematics and statistics for hundreds of years. It is well understood, and the essential methods are well documented. There are many more approaches to probability than that described here. Probability is a subject that is on the syllabus of almost every science course at degree level and so it pervades the understanding of uncertainty through many disciplines. It is not, however, the only way to treat uncertainty.

3.4. Vagueness

In contrast with error and probability, which are steeped in the mathematical and statistical literature, vagueness is in the realm of philosophy and logic and has been described as one of the fundamental challenges to those disciplines [WIL 94]. It is relatively easy to show that a concept is "vague", and the classic pedagogic exposition uses the case of the "bald" man. If a person with no hair at all is considered bald, then is a person with one hair bald? Usually, and using any working definition of "bald", the answer to this would be "yes". If a person with one hair is bald then is a person with two hairs bald? Again, yes. If you continue the argument, one hair at a time, then the addition of a single hair never turns a bald man into a man with a full head of hair. On the other hand, you would be very uncomfortable stating that someone with plenty of hair was bald because this is illogical [BUR 92]. This is known as the Sorites Paradox which, little by little, can present the reverse logical argument that someone with plenty of hair is bald! A number of resolutions to the paradox have been suggested, but the most widely accepted is that the logic employed permits only a Boolean response ("yes" or "no") to the question. A graded response is not acceptable; and yet, there is a degree to which a person can be bald. It is also possible that the initial question is false, because bald would normally be qualified if we were examining it in detail; so we might ask whether someone was "completely bald", and we might define that as someone with no hair at all. Can we ever be certain that a person has absolutely no hair on their head? Furthermore, where on their neck and face is the limit of the head so that we can judge whether there is any hair on it? You are eventually forced to admit that by logical argument, it is impossible to specify whether someone is "completely", "absolutely", "partially", or "not at all" bald, given a count of hairs on their head, even if the count is absolutely correct. So no matter the precision of the measurement, the allocation of any man to the set of bald men is inherently vague.

The Sorites Paradox is one method that is commonly used to define vague concepts. If a concept is Sorites susceptible, it is vague. Very many geographical phenomena are Sorites susceptible, including concepts and objects from the natural and built environments. When, exactly, is a house a house; a settlement, a settlement; a city a city; a podzol, a podzol; an oak woodland, an oak woodland? The questions always revolve around the threshold value of some measurable parameter or the opinion of some individual, expert or otherwise.

Fuzzy-set theory was introduced by Zadeh [ZAD 65] as an alternative to Boolean sets. Membership of an object in a Boolean set is absolute, that is, it either belongs or it does not belong, and membership is defined by one of two integer values {0,1}. In contrast, membership of a fuzzy set is defined by a real number in the range [0,1] (the change in type of brackets indicates the real and integer nature of the number range). Membership or non-membership of the set is identified by the terminal values, while all intervening values define an intermediate degree of belonging to the set, so that, for example, a membership of 0.25 reflects a smaller degree of belonging to the set than a membership 0.5. The object described is less like the central concept of the set.

Fuzzy memberships are commonly identified by one of two methods [ROB 88]:

– The *similarity relation model* is data driven and involves searching for pattern within a dataset and is similar to traditional clustering and classification methods; the most widespread methods are the Fuzzy c Means algorithm [BEZ 81] and fuzzy neural networks [FOO 96].

– The *semantic import model*, in contrast, is derived from a formula or formulae specified by the user or another expert [ALT 94; WAN 90].

Many studies have applied fuzzy-set theory to geographical information processing. There are several good introductions to the application of fuzzy sets in geographical data processing, including books by Leung [LEU 88] and Burrough and Frank [BUR 96]. Fuzzy-set theory is now only one of an increasing number of soft set theories [PAW 82; AHL 03].

3.5. Ambiguity

The concepts and consequences of ambiguity (Figure 3.1) in geographical information are the subject of an increasing amount of research. Ambiguity occurs when there is doubt as to how a phenomenon should be classified because of differing perceptions of that phenomenon. Two types of ambiguity have been recognized, namely discord and non-specificity.

3.5.1. *Discord*

The conflicting territorial claims of nation states over a single piece of land is the most obvious form of discord. History is filled with this type of ambiguity, and the discord which results. Kashmir (between India and Pakistan) and the neighboring Himalayan mountains (between China and India) are just two examples in the modern world involving border conflicts and disagreements. Similarly, the existence or non-existence of a nation of Kurds is another source of discord. All represent mismatches between the political geography of the nation states, and the aspirations of people [PRE 87].

As has already been noted, many if not most phenomena in the natural environment are also ill-defined. The inherent complexity in the defining of soil, for example, is revealed by the fact that many countries have slightly different definitions of what constitutes a soil at all (see [AVE 80; SOI 75]), as well as by the complexity and the volume of literature on defining the spatial and attribute boundaries between soil types [LAG 96]. Furthermore, no two national classification schemes have either the same names for soils or the same definitions if they happen to share names, which means that many soil profiles are assigned to different classes in different schemes [SOI 75].

There is an increasing realization that discord is endemic in land cover classification as well. At the local level, consider the area shown in Figure 3.3, where four experienced air photo interpreters, using the same data, the same methodology, and the same formal class descriptions, all delineate the same scene in different ways. As with soil, within a single classification this is not a problem; the ambiguity arises where there are multiple mappings of the same area, whether these mappings take place once or on multiple occasions [COM 03]. Land cover mapping has the additional problem that it is frequently mapped as a surrogate for land use mapping. This is not necessarily a problem for the technicians performing the classification, but can be a major problem for those using land cover information in policy-making [FIS 05] and, in the case of soil information, for those attempting to produce regional or global soil maps from mapping by national agencies, such as the soil map of the EC or the world [FAO 90].

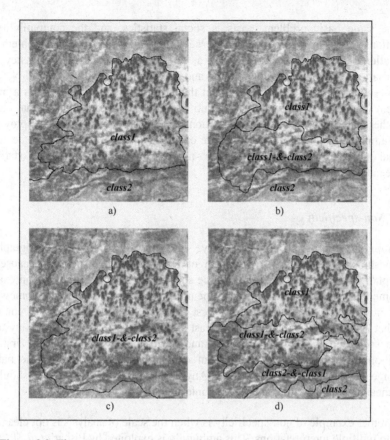

Figure 3.3. *The same area may be mapped differently by individual interpreters, as illustrated by the boundary placements: a) at the point at which there is more **class1** than **class2**; b) the "zone of transition" between classes 1 and 2 is represented by a mosaic of **class1-&-class2**; c) the whole area is allocated into a **class1-&-class2** mosaic; d) the two distinct areas of **class1** and **class2** are separated by two mosaics of **class1-&-class2** and **class2-&-class1** [COM 02]*

A number of solutions have been suggested for the problem of discord. All are based on an expert judgment as to the compatibility between classifications. Thus Comber *et al.* [COM 04a], for example, use expert look-up tables as well as producer-supplied metadata to compare the 1990 and 2000 classifications of land cover in Britain, and Ahlqvist *et al.* [AHL 00; AHL 03] use personal (expert) judgment to compare vegetation change over a longer period. The former work is grounded firmly in artificial intelligence while the latter makes extensive use of rough and fuzzy sets to accommodate the uncertainty in the correspondence of classes.

The same basic problem exists is social statistics, and the compatibility of definitions used in the collection of such statistics is a major problem for international statistical agencies, such as the European Statistical Agency, the Organisation for Economic Cooperation and Development, and the United Nations. The fact that different countries conduct their census in different years is a minor problem compared to the different definitions used in the collection of that census and other statistical information. Definitions of long-term illness or unemployment are examples of social categories that are open to manipulation by governments for internal political reasons, but which make reporting international statistics problematic.

3.5.2. *Non-specificity*

Ambiguity through non-specificity can be illustrated by geographical relationships. The relation "A is north of B" is itself non-specific because the concept "north of" can have at least three specific meanings: that A lies on exactly the same line of longitude and towards the north pole from B; that A lies somewhere to the north of a line running east to west through B; or, the meaning in common use, that A lies between perhaps north-east and north-west, but is most likely to lie in the sector between north-north-east and north-north-west of B. The first two definitions are precise and specific, but equally valid. The third is the natural language concept, which is itself vague. Any lack of definition as to which should be used means that uncertainty arises in the interpretation of "north of".

Another example of non-specificity is when the scale of analysis is not clear or is open to multiple interpretations. This ambiguity is exploited by Fisher *et al.* [FIS 04] in their analysis of peakness in the British Lake District, and in their identification of the footprints of mountains using fuzzy-set theory, while Comber *et al.*, in Chapter 7 in this book, use endorsement theory to address the problem of non-specificity in land cover information.

3.6. Data quality

Data quality is a concept related to uncertainty. In its present form it has been conceptualized by the producers of spatial data [FGD 98]. It mimics the concerns of institutional producers of cartographic products, but is informed by the need to transfer spatial information. The principal organizations specifying transfer standards are either industrial or institutional, and the process is dominated by two sectors: the national mapping agencies and the software companies. The process is now advanced by the Open GIS Consortium [BUE 98], and has been further developed in the areas of ISO 19000 [DAS 02]. There are six principal areas of data quality listed in Table 3.2. Four are widely recognized [FGD 98], and two others are suggested in chapters in the book edited by Guptill and Morrison [GUP 95], and are discussed by Servigne *et al.* (Chapter 10).

	Lineage	
	Accuracy	Positional
		Attribute
Data Quality	Completeness	
	Logical Consistency	
	Semantic Accuracy	
	Currency	

Table 3.2. *Some recognized aspects of data quality [GUP 95]*

It is remarkable that most of the research in uncertainty has had very little influence on the development of the terms used to describe quality. Work on error is certainly embodied in the error measures. Concern with either spatial autocorrelation in the description of error, or the idea that error in fields is not a global value, is not addressed by methods for describing error outlined within the standards. Object-based error can be defined in terms of the standards, however, and does go some way toward addressing this concern for object-oriented databases. The possibility that the information may be vague or that the classification scheme may be discordant with another classification scheme is not addressed.

3.7. Precision

One other term used in describing data quality is precision. This refers to the accuracy with which a measurement can be made, recorded, or calculated. Current laser ranging systems (LiDAR) used in measuring the height of the surface of the earth have a precision in the order of 5 cm, which means that they can record the elevation at a point to the nearest 5 cm. Values may be reported to higher precision, but they intrinsically only have this precision. Similarly, databases can have a precision. More traditional digital elevation models (DEM) have typically been distributed as integer values reporting meters. The precision of such a DEM is to the nearest meter. No measurement less than a meter has any meaning because it is not recorded in the database, which of course also means that whatever value is recorded for a location, the actual value may be ±1 m of that value, before error is taken into account. Similarly, programming languages all work to a precision in terms of the number of significant figures and the size of number that can be determined. Thus Java, for example, can do single (known as float) and double precision calculations using 32 and 64 bits to store each number, allowing values to be determined in the range $\pm 3.4 * 10^{38}$ to 6 decimal places and $\pm 1.8 * 10^{308}$ to 15 decimal places, respectively. In other words, no value that needs to be known to more than 15 digits can ever be calculated, except with specific algorithms.

3.8. Conclusion: uncertainty in practice

A variety of causes of, and approaches, to uncertainty are endemic within geographical information. Anyone using uncertain information (that is, the overwhelming majority of GIS users) needs to think carefully about the possible sources of uncertainty and how they may be addressed. At the heart of it all is the conceptualization of the nature of the information. Any analysis which does not accommodate data uncertainty (both error and vagueness) can limit the usefulness of that analysis. On the other hand, an appropriate conceptualization of uncertainty and the application of related procedures creates a rich analytical environment where decision-making based on spatial information is facilitated, not only by objective orderings of alternatives, but also by giving confidence in relation to those alternatives.

It is notable that the standards for data quality are not specifically informed by the research in uncertainty. This is in part due to the conceptualization of most spatial information developed by the national mapping agencies (and other producers) who work within the Boolean map paradigm of production cartography [FIS 98b] and so are content with probabilistic concepts of accuracy and consistency. When data producers reconceptualize their information to accommodate the diversity of modern models of uncertainty, a wider set of standards will be required, which will necessarily be informed by the research in uncertainty.

3.9. References

[AHL 00] AHLQVIST O., KEUKELAAR J. AND OUKBIR K., "Rough classification and accuracy assessment", *International Journal of Geographical Information Science*, 2000, vol 14, p 475–96.

[AHL 03] AHLQVIST O., KEUKELAAR J. ANDBIR K., "Rough and fuzzy geographical data integration", *International Journal of Geographical Information Science*, 2003, vol 17, p 223–34.

[ALT 94] ALTMAN D., "Fuzzy set theoretic approaches for handling imprecision in spatial analysis", *International Journal of Geographical Information Systems*, 1994, vol 8, p 271–89.

[AVE 80] AVERY B.W., *Soil Classification for England and Wales (Higher Categories)*, Soil Survey Technical Monograph 14, 1980, Harpenden.

[BEZ 81] BEZDEK J.C., *Pattern Recognition with Fuzzy Objective Function Algorithms*, 1981, New York, Plenum Press.

[BUE 98] BUEHLER K. and MCKEE L. (eds), *The Open GIS Guide: Introduction to Interoperable Geoprocessing and the OpenGIS Specification*, 3rd edn, Open GIS Technical Committee, Wayland MA, Draft Specification, http://www.opengis.org/techno/guide/ guide980615/Guide980601.rtf, 1998.

[BUR 92] BURROUGH P.A. "Are GIS data structures too simple minded?", *Computers & Geosciences*, 1992, vol 18, p 395–400.

[BUR 96] BURROUGH P.A. and FRANK A. (eds), *Spatial Conceptual Models for Geographic Objects with Undetermined Boundaries*, 1996, London, Taylor & Francis.

[COM 02] COMBER A.J., Automated land cover change, PhD Thesis, University of Aberdeen, 2002.

[COM 03] COMBER A.J., FISHER P. and WADSWORTH R., "Actor Network Theory: a suitable framework to understand how land cover mapping projects develop?", *Land Use Policy*, 2003, vol 20, p 299–309.

[COM 04a] COMBER A.J., FISHER P. and WADSWORTH R., "Integrating land cover data with different ontologies: identifying change from inconsistency", *International Journal of Geographical Information Science*, 2004, vol 18, p 691–708.

[COM 04b] COMBER A.J., FISHER P. and WADSWORTH R., "What is land cover?", *Environment and Planning B, Planning and Design*, 2004, vol 32, p 199–209.

[CON 83] CONGALTON R.G. and MEAD R.A., "A quantitative method to test for consistency and correctness in photointerpretation", *Photogrammetric Engineering and Remote Sensing*, 1983, vol 49, p 69–74.

[DAS 02] DASSONVILLE L., VAUGLIN F., JAKOBSSON A. and LUZET C., "Quality management, data quality and users, metadata for geographical information", in *Spatial Data Quality*, Shi W., Fisher P.F. and Goodchild M.F. (eds), 2002, London, Taylor & Francis, p 202–15.

[EDW 94] EDWARDS G., "Characteristics and maintaining polygons with fuzzy boundaries in geographic information systems", in *Advances in GIS Research: Proceedings of the Sixth International Symposium on Spatial Data Handling*, Waugh T.C. and Healey R.G. (eds), 1994, London, Taylor & Francis, p 223–39.

[FAO 90] FAO/UNESCO, *"Soil Map of the World: Revised Legend World Soil Resources"*, Report 60, 1990, FAO, Rome.

[FGD 98] FEDERAL GEOGRAPHIC DATA COMMITTEE, *"Content Standard for Digital Geospatial Metadata"*, 1998, FGDC-STD-001-1998, Springfield, Virginia, National Technical Information Service, Computer Products Office.

[FIS 98a] FISHER P.F., "Improved modelling of elevation error with geostatistics", *GeoInformatica*, 1998, vol 2, p 215–33.

[FIS 98b] FISHER P.F., "Is GIS hidebound by the legacy of cartography?", *Cartographic Journal*, 1998, vol 35, p 5–9.

[FIS 99] FISHER P.F., "Models of Uncertainty in Spatial Data", in *Geographical Information Systems: Principles, Techniques, Management and Applications*, Longley P., Goodchild, M. Maguire D. and Rhind D. (eds), 1999, vol 1, New York, john Wiley & Sons, p 191–205.

[FIS 04] FISHER P.F., CHENG T. and WOOD J., "Where is Helvellyn? Multiscale morphometry and the mountains of the English Lake District", *Transactions of the Institute of British Geographers*, 2004, vol 29, p 106–28.

[FIS 05] FISHER P.F., COMBER A.J. and WADSWORTH R., "Land use and land cover: contradiction or complement?", in *Re-presenting GIS*, Unwin D.J. and Fisher P.F. (eds), 2005, Wiley, London.

[FOO 96] FOODY G. M., "Approaches to the production and evaluation of fuzzy land cover classification from remotely-sensed data", *International Journal of Remote Sensing*, 1996, vol 17, p 1317–40.

[GUP 95] GUPTILL S.C. and MORRISON J.L. (eds), *Elements of Spatial Data Quality*, 1995, Oxford, Elsevier.

[HEU 99] HEUVELINK G.B.M., "Propagation of error in spatial modelling with GIS", in *Geographical Information Systems: Principles, Techniques, Management and Applications*, Longley P., Goodchild M., Maguire D. and Rhind D. (eds), 1999, vol 1, New York, John Wiley & Sons, p 207–17.

[KLI 95] KLIR G.J. and YUAN B., *Fuzzy Sets and Fuzzy Logic: Theory and Applications*, 1995, Englewood Cliff, Prentice Hall.

[LAG 96] LAGACHERIE P., ANDRIEUX P. and BOUZIGUES R., "The soil boundaries: from reality to coding in GIS", in *Spatial Conceptual Models for Geographic Objects with Undetermined Boundaries*, Burrough P.A and Frank A. (eds), 1996, London, Taylor & Francis, p 275–86.

[LAN 92] LANTER D.P. and VEREGIN H., "A research paradigm for propagating error in layer-based GIS", *Photogrammetric Engineering and Remote Sensing*, 1992, vol 58, p 825–33.

[LEU 88] LEUNG Y.C., *Spatial Analysis and Planning under Imprecision*, 1988, New York, Elsevier.

[LUN 04] LUND H.G., *Definitions of Forest, Deforestation, Afforestation, and Reforestation*, 2004, http://home.comcast.net/~gyde/DEFpaper.htm. [Available 05/05/04].

[MAR 99] MARTIN D., "Spatial representation: the social scientist's perspective", in *Geographical Information Systems: Principles, Techniques, Management and Applications*, Longley P., Goodchild M., Maguire D. and Rhind D. (eds), 1999, vol 1, New York, John Wiley & Sons, p 71–80.

[MOR 93] MORACZEWSKI I.R., "Fuzzy logic for phytosociology 1: syntaxa as vague concepts", *Vegetatio*, 1993, vol 106, p. 1–11.

[OFF 97] OFFICE OF NATIONAL STATISTICS, *Harmonized Concepts and Questions for Government Social Surveys*, 1997, London, Her Majesty's Stationary Office.

[OPE 95] OPENSHAW S. (ed), "Census user's handbook", *GeoInformation International*, 1995, Cambridge, GeInformation International.

[PAW 82] PAWLAK Z., "Rough Sets", *International Journal of Computer and Information Sciences*, 1982, vol 11, p 341–56.

[PRE 87] PRESCOTT J.R.V., *Political Frontiers and Boundaries*, 1987, London, Allen and Unwin.

[ROB 88] ROBINSON V.B., "Some implications of fuzzy set theory applied to geographic databases", *Computers, Environment and Urban Systems*, 1988, vol 12, p 89–98.

[SMI 01] SMITH B., "Fiat Objects", *Topoi*, 2001, vol 20, p 131–48.

[SOI 75] SOIL SURVEY STAFF, *Soil Taxonomy: A Basic System of Soil Classification for Making and Interpreting Soil Surveys*, 1975, USDA Agricultural Handbook 436, Washington DC, Government Printing Office.

[TAY 82] TAYLOR J.R., *An Introduction to Error Analysis: The Study of Uncertainties in Physical Measurements*, 1982, Oxford, Oxford University Press.

[THU 00] THURSTAIN-GOODWIN M. and UNWIN D., "Defining and Delineating the Central Areas of Towns for Statistical Monitoring using Continuous Surface Representations", *Transactions in GIS*, 2000, vol 4, p 305–18.

[WAN 90] WANG F., HALL G.B. and SUBARYONO, "Fuzzy information representation and processing in conventional GIS software: database design and application", *International Journal of Geographical Information Systems*, 1990. vol 4, p 261–83.

[WIL 94] WILLIAMSON T., *Vagueness*, 1994, London, Routledge.

[ZAD 65] ZADEH L.A., "Fuzzy sets", *Information and Control*, 1965, vol 8, p 338–53.

Chapter 4

Quality of Raster Data

4.1. Introduction

The quality of raster data measures the ability of each pixel to provide users with reliable spectral and thematic information, together with a known Earth location.

This quality is a necessary condition that will enable:

– processing large areas, seamlessly assembling data acquired at different periods of time (mosaics, synthesis);

– monitoring geophysical measurements over time (multi-temporal analysis);

– merging data acquired from different instruments to improve the data analysis and classification deriving new thematic information with better accuracy.

Quality of raster data is processed in two distinct domains:

– geometric qualification; and

– radiometric qualification.

Activities performed in these two domains involve different methods and are generally undertaken by experts who have specific knowledge: cartography and geodesy for geometry; radiative transfer physics and thematic skills for radiometry.

To enable radiometric qualification, for example when operating validation from measurements of another instrument, it is first required to "geolocate" (also referred

Chapter written by Serge RIAZANOFF (the "geometry quality" section) and Richard SANTER (the "radiometry quality" section).

as "collocate") Earth observation data. This is the reason why the geometry qualification must be performed *before* the radiometry qualification.

4.2. Geometry quality

Geometric processing performed on one image transforms it, and makes it conform to a cartographic reference system. The product of such a transformation is called a *spacemap*.

Figure 4.1. *Orthorectified aerial photograph superposed on a map*
(Courtesy TTI Productions)

Many factors impact on the quality of the acquired image; these are presented in this section.

After having provided some definitions and having set a theoretical framework, various geometry defects will be presented, complemented by the presentation of the two methods most frequently used to verify image geometry.

4.2.1. *Image reference system and modeling of the viewing geometry*

4.2.1.1. *Image reference system in matrix representation*

One image is made up of m lines, each one containing n points (or pixels), and is generally stored in one or more files the first point of which matches the upper-left corner of the image. In order to comply with this constraint and to handle a direct base (according to a mathematical point of view), the use of a Cartesian reference system, giving first the coordinate l of the line (see Figure 4.2), is recommended.

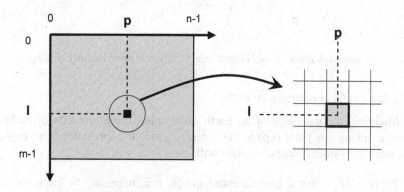

Figure 4.2. *Image reference system and elementary grid-cell*

Point (l,p) of an image has a footprint on the ground which is represented by a square. Such a footprint defines an elementary grid-cell (see its definition in the next section) that may be represented on the Earth's surface, and for which may be provided geodetic coordinates. Common usage is to consider these coordinates as at the center of the grid-cell. Nevertheless, some geocoded images consider that the geodetic coordinates refer to the upper-left corner of the grid-cell.

Any uncertainty regarding the location within the grid-cell (center or corner) of the geodetic coordinates could lead to a systematic error of 0.5 pixels along the two directions (that is, a total of $\sqrt{2}/2$ pixels).

Note also that a square footprint representation is intrinsically inadequate because the radiometry measured at one point in fact corresponds to the integration of backscattered energies across a surface (called "point spread distribution") which almost never matches a square pattern (see Figure 4.3).

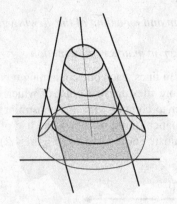

Figure 4.3. *Point Spread Distribution (PSD) over an elementary grid-cell*

4.2.1.2. *Direct and inverse localization*

Modeling the geometry of an Earth observation image consists of defining a relation linking any point (l,p) of the image to geodetic coordinates (λ,φ) expressed in a reference system attached to the Earth.[1]

To be able to use a georeferenced image, it is necessary to know the *direct localization function* f (more generally called *direct deformation* model) and/or the *inverse localization function* f^{-1} (also called *inverse deformation model*), which are mathematical formulae given in the generic form:

direct localization function $f\,(l, p, h, var) = (\lambda, \varphi)$ [4.1]

inverse localization function $f^{-1}(\lambda, \varphi, h, var) = (l, p)$ [4.2]

where:

(l,p)	are the coordinates of the point in the image,
(λ, φ)	are the geodetic coordinates of this point on the earth's surface,
h	matches the altitude of the point above the earth's surface,
var	are auxiliary variables, private resources, required by the model (for example the coordinates of the viewing vector, the resolution, etc.).

---

1 In this chapter, (λ,φ) denotes the geodetic coordinates that usually match the (longitude, latitude) couple without considering the case of coordinates (X,Y) expressed in other geodetic reference systems, for example the Lambert or UTM systems. Conversion between these coordinate systems is performed using standard formulae of projection change, the description of which is outside the scope of this book.

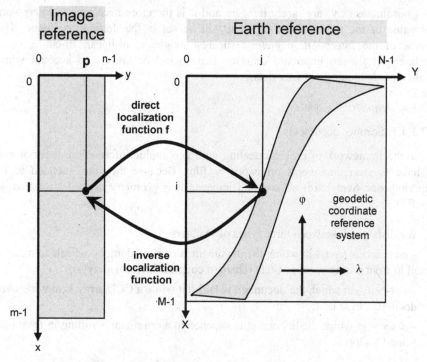

Figure 4.4. *Image and terrestrial reference frames*

4.2.1.3. *Geometric transforms of images*

As seen previously, the *direct localization function* may be used to determine the point location in a terrestrial reference frame, but may also be used, combined with the *inverse localization function*, to transform the image geometry (we sometimes say *to* (*re*)*project*) into a terrestrial frame, producing a spacemap in a classical projection (see Figure 4.4).

The *direct localization function* makes it possible to compute the projection of the four corners of the original image (or all the points of the image boundary when the transformation is complex, as shown in Figure 4.4) and determine the size and the terrestrial location (footprint) of the destination image.

The *inverse location function* plays an essential role while reprojecting the image because it enables the construction of the destination image point by point, retrieving for each point (i,j) of this destination image the coordinates (x,y) of the antecedent point in the original image.

Coordinates (x,y) are rarely integers and it is therefore necessary to interpolate the value of the radiometry R(i,j) that will be set in the destination image. The choice of the *interpolation method* (nearest neighbor, bi-linear, bi-cubic, sinus cardinal, etc.) is an important criterion that should be taken into account while evaluating the quality of raster data.

4.2.1.4. *Acquisition models*

4.2.1.4.1. Scanner documents

In the framework of projects dealing with geographic information, it is not rare to have to scan documents on paper or film. Because they are intended to be superimposed over Earth observation images, their geometry quality must also be verified.

We distinguish between three types of scanners:

– *continuous feed* – in which the document is carried along by wheels to make it scroll in front of one or more CCD (charged coupled devices) array(s);

– *flat-bed* – in which the document is laid flat while a CCD array is moving over the document to scan it;

– *drum* – in which the document is attached to a drum that is rolling in front of a laser and CCD(s).

These three technologies enable the processing of documents of different sizes, and produce digital images of very different qualities. *Drum-scanners* are hardware that are able to reach the highest resolution and have a high geometric quality, but are expensive. *Continuous feed scanners* are able to process large documents, are inexpensive, but often produce deformations in the output image due to the friction of the wheels (see one example in Figure 4.8). *Flat-bed scanners*, except very expensive ones, are only able to process small documents (A4 or A3).

4.2.1.4.2. Viewing models

There are at least four basic types of viewing geometry from aerial or space platforms (see Figure 4.5):

– *conic* – matches a one-shoot photograph of the terrestrial surface (this type of geometry is the one of most usual instruments onboard airplanes);

– *push-broom* – in which a line of terrain across the flight line (also called NADIR) is seen with a constant scanning frequency (for example, SPOT-HRV);

– *scanner* – in which an oscillating CCD array scans the terrain from one side of the NADIR to the other side (for example, Landsat TM);

– *radar* – which analyzes the go-return time and the phases of an emitted signal to one side of the NADIR.

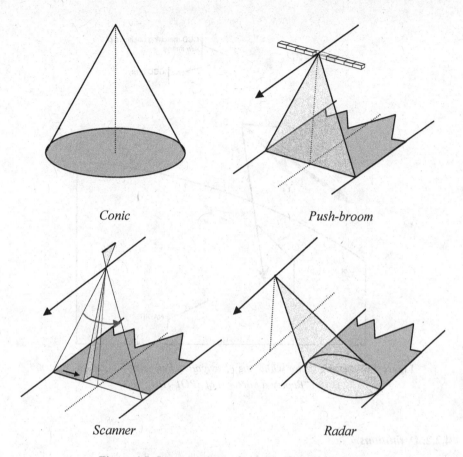

Conic   *Push-broom*

Scanner   *Radar*

Figure 4.5. *Basic viewing technologies from platforms*

This list of viewing models is not exhaustive. Other technologies exist, often more complex, such as the "TV-scanning" of Meteosat satellites or the "conic scanning" of the ATSR instrument onboard ERS and Envisat satellites, etc.

Each technology leads to its own geometry defects, which are presented in section 4.1.2.3.

Direct and/or inverse localization functions may be complex. They are the expression of a viewing model involving the position of the platform, its instantaneous attitude (that is, its orientation), the geometry of the instrument, and an Earth model that may possibly include an elevation model.

For example, Figure 4.6 illustrates one of the various reference frames used to model the acquisition of the HRV instrument onboard SPOT satellites.

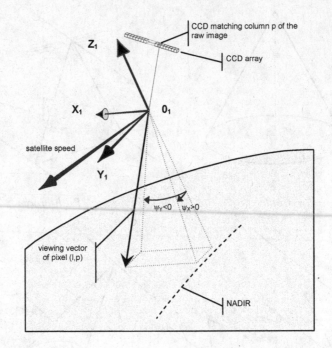

Figure 4.6. *Viewing vector within the "Navigation Reference Coordinate System"*
("Repère à piloter") of SPOT-HRV

4.2.2. *Definitions*

We may notice great differences in the meaning of terms used in image processing activities and, in particular, in relation to the quality control of images. The use of both French and English has increased such differences. Although not claiming to be an exhaustive list, the definitions set out in this section are an attempt to clarify and to complement the definitions found in the dictionary [CON 97].

4.2.2.1. *Georeferenced image*

A *georeferenced image* is an image in which any point (l,p), geodetic coordinates (λ,φ) or cartographic coordinates (X,Y) are given by an analytical formula or by one algorithm (computer program) that may be complex (for example, the level 1A image of SPOT in the opposite figure).

The direct localization function (equation 4.1) assumes this role; but other localization functions may exist for the same image. For example, ancillary data of SPOT scenes provided with polynomial coefficients making it possible to approximate this localization function (for example, the level 1B image of SPOT in the opposite figure).

4.2.2.2. *Geocoded image*

We usually define a *geocoded image* as an image for which a linear direct localization function exists giving the geodetic coordinates (λ, φ) or the cartographic (X,Y) for any point (l,p) of the image.

$$\begin{cases} X = X_0 + P_w \times p \\ Y = Y_0 - P_h \times l \end{cases} \qquad [4.3]$$

where:

(l,p)	are the coordinates of the point in the image,
(X,Y)	are the cartographic coordinates or geodetic coordinates (λ, φ),
(X_0,Y_0)	are the coordinates of the upper-left corner,
P_w and P_h	are the on-ground pixel width and height respectively.

4.2.2.3. *Orthorectified image*

An *orthorectified image* is an image for which parallax effects, which exist due to the relief and to the viewing vector (perspective effect), have been corrected. Orthorectification requires using a digital elevation model (DEM). An orthorectified image is a geocoded image with cartographic accuracy.

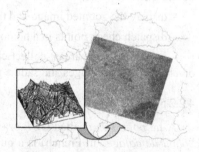

4.2.2.4. *Check points*

A *check point* ("*point de contrôle*") is a quadruplet (l,p,Λ,ϑ), for which the geodetic coordinates (Λ,ϑ) have been provided by an external cartographic reference: map, geocoded image acting as reference, GPS survey, geodetic point, etc. A check point is used to assess the localization accuracy of the image to be checked.

p   ϑ

I   Λ

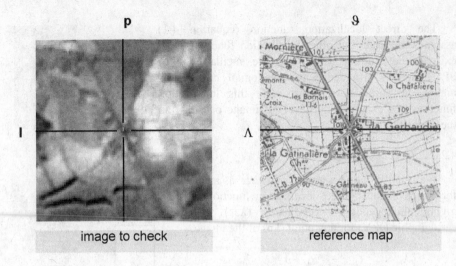

image to check   reference map

Figure 4.7. *Check point, at a crossroads, extracted from a map*

An accurate extraction of check points is a difficult exercise that depends on the scale of the documents, the type of objects found in the landscape (for example, a crossroads is often more precise than a pier), the precision of the tool used to pick the point, the operator skills, etc.

Nevertheless, we can enumerate some basic rules for the survey of check points:

– favor intersections having nearly perpendicular axes (roads, field corners, building corners, etc.);

– display the zoomed image and the reference document at the same scale;

– dispatch check points in a homogeneous way within the image to be checked;

– in mountainous areas, collect check points, both at the bottom of valleys and on elevated zones (ridges and saddle points).

4.2.2.5. *Tie points*

A *tie point* (also called *GCP* or *control point* in English and "*point d'appui*" or "*point de calage*" in French) is a quadruplet (l,p,Λ,ϑ) extracted like a *check point*, but that is used to register the direct localization function, solving uncertainties linked to the unknowns of its attached model.

For example, if the location (X_S,Y_S,Z_S) of the platform is not known when the terrain has been imaged, at least three points will be required to interpolate this location using a technique called *triangulation*. By using more tie points, we may refine this platform location minimizing the mean quadratic error.

We may often notice confusion between *check points* and *tie points*. This confusion is certainly due to the double semantic of the word "control", which means both "to assess the quality" and "to monitor the model". In French, the first meaning is more often used while the second one seems to be favored in English.

4.2.2.6. *Localization error*

The *localization error* of one point is the distance between the geodetic location (λ, φ) given by the direct localization (equation 4.1) and the location (Λ, ϑ) observed in the surveyed check point.

$$e \;=\; dist\big[(\lambda, \varphi), (\Lambda, \theta)\big] \tag{4.4}$$

where:

    dist is the distance function to be defined: Euclidian distance or geodetic distance, etc.

4.2.2.7. *Mean quadratic error*

The *mean quadratic error*, also called root mean square (RMS) error, is a mean of the localization errors.

$$rmse \;=\; \sqrt{\frac{1}{n}\sum_{i=1}^{n} e_i^{2}} \;=\; \sqrt{\frac{1}{n}\sum_{i=1}^{n} dist\big[(\lambda_i, \varphi_i), (\Lambda_i, \theta_i)\big]^2} \tag{4.5}$$

where:

    n is the number of check points,

    e_i is the localization error of point i (i=1 ... n),

    (λ_i, φ_i) is the geodetic location computed by the direct localization function applied to point i (i=1 ... n),

    (Λ_i, ϑ_i) is the geodetic location given by the reference cartographic document at point i (i=1 ... n),

    dist is the distance function to be defined: Euclidian distance or geodetic distance, etc.

4.2.2.8. *Error vector field*

The *error vector* of one point i joins the location (λ_i, φ_i) predicted by the direct localization function (corrupted by an error) up to the (Λ_i, ϑ_i) provided by the reference document (which is supposed to be the correct location).

$$\vec{v_i} \;=\; \begin{pmatrix} \Lambda_i - \lambda_i \\ \theta_i - \varphi_i \end{pmatrix} \tag{4.6}$$

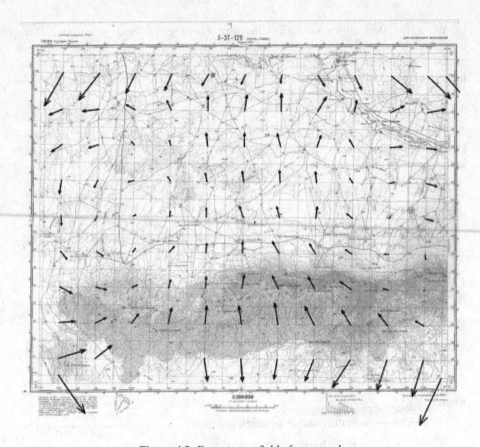

Figure 4.8. *Error vector field of a scanned map*

Displaying the error vector field is an excellent way to understand the defects of the direct localization function. For example, Figure 4.8 illustrates the case of a paper map having been scanned by a continuous feed scanner, the wheels of which have caused extensions/compressions of the paper.

4.2.2.9. Native projection of a map

The *native projection of a map* is the reference coordinate system that has been used for its generation. Horizontal (X) and vertical (Y) axes of the native projection are strictly parallel to the map borders and graduations are spaced with a constant interval.

A map may include grids, tick marks, cross marks matching another projection, and sometimes many other projections.

When a map is scanned for geocoding, it is recommended that one collects a significant number of tie points (see, for example, Figure 4.8), regularly spaced and distributed within the image, and then transforms this scanned image in its native projection using a polynomial interpolation of degree 2 and maximum 3 (high degrees may lead to oscillations).

4.2.3. Some geometry defects

4.2.3.1. Absolute localization defect

The algebraic mean of each component in latitude and longitude makes it possible to detect a translation between the location (λ_i, φ_i) predicted by the direct localization function and the location (Λ_i, ϑ_i) given by the reference cartographic documents.

$$
\begin{cases}
\overline{m}_\lambda &= \dfrac{1}{n}\sum_{i=1}^{n}(\Lambda_i - \lambda_i) \\
\overline{m}_\varphi &= \dfrac{1}{n}\sum_{i=1}^{n}(\theta_i - \varphi_i)
\end{cases}
\qquad [4.7]
$$

This vector (m_λ, m_φ) measures the *absolute location error*.

For images acquired by scrolling satellites, this defect is generally due to erroneous values of the platform ephemeris (giving the location and speed) and, in particular, in their attached time-stamp (date/time).

For example, the opposite figure shows the location on Earth of a SPOT scene acquired by a lateral viewing and the terrestrial projection of the eight ephemeris data contained in the product header. A time-stamp error of these data would lead one to move the location of the SPOT scene along a line parallel to the NADIR.

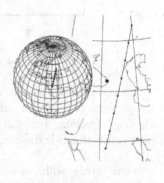

The absolute location error may easily be corrected by just adding values m_λ and m_φ to the coordinates of the upper-left corner that are generally kept as auxiliary data of a geocoded image in almost all the geographic information systems (GIS).

4.2.3.2. Global defects of internal geometry

Analysis of error vector directions may enable quick detection of global deformations (called low-frequency deformations) of an image.

The absolute localization error seen in the previous section is also a global deformation in which error vectors are almost parallel (Figure 4.9a).

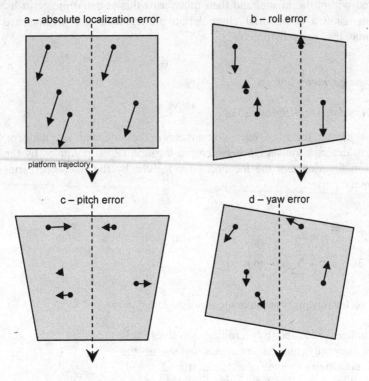

Figure 4.9. *Some global defects of internal geometry*

Defects of roll (b), pitch (c) and yaw (d) are due to uncertainties in the attitude of the platform (airborne or satellite) and therefore there are uncertainties about the viewing direction that will intersect the Earth. Roll and pitch errors lead to perspective effects in the image, while a yaw error leads to a rotation of the image.

Over areas with small relief variations, these defects may be corrected individually and very locally by using low-degree polynomial transforms.

In all the other cases, an accurate orthorectification will integrate ephemeris data (location and speed) of the platform and its attitude data (roll, pitch, and yaw angles around a reference frame linked to the platform). A time shall be attached to these data which will enable the establishment of a relation to the time of the image line (in the case of a push-broom or a scanner viewing model) or to the instantaneous time of acquisition (in the case of a conic viewing model).

4.2.3.3. *Local defects of internal geometry*

Local defects of internal geometry (called high-frequency deformations) may have different causes. Some of the most usual ones are enumerated below.

4.2.3.3.1. DEM accuracy

When an image is orthorectified, parallax defects (see Figure 4.10) that combine effects of the elevation α of the viewing vector with the elevation h of the terrain will be corrected.

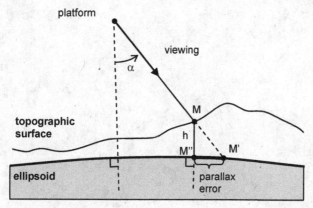

Figure 4.10. *Parallax error*

To be accurate, the parallax error is the geodetic distance (M',M''), measured along the ellipsoid, between the vertical projection M'' of point M on the ellipsoid and the intersection M' of the viewing vector with this ellipsoid. Nevertheless, this error may also be approximated using the simple formula:

$$\overline{M'M'} \;=\; h \times \tan(\alpha) \qquad\qquad [4.8]$$

To be consistent, the viewing model and the DEM will make reference to the same terrestrial representation: The WGS84 ellipsoid is generally used. In particular, the DEM will contain elevations with reference to the ellipsoid and not to altitudes above the geoid that may differ from the WGS84 by up to 200 meters.

The quality of a DEM is defined by two measurements:

– *planimetric resolution* – which gives the on-ground spacing between two consecutive elevation values; and

– *elevation resolution (or vertical resolution)* – which defines the vertical accuracy of elevation measurements (generally measured in meters or decimeters).

In orthorectified images, when the error vector direction depends on the elevation of the check points (see Figure 4.11), local defects of the internal geometry may be due to the accuracy of the DEM that is being used, to an error in its geographic location, or to the interpolation method being used to estimate elevations in the DEM.

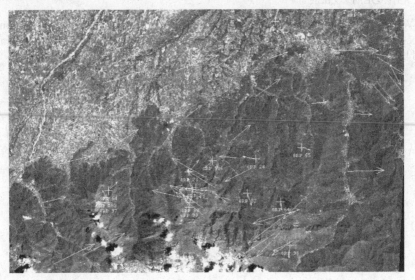

Figure 4.11. *Defects of parallax corrections in a mountainous area*

4.2.3.3.2. Defects due to the instrument

Push-broom instrument

In a push-broom instrument, CCDs are stored in one or more arrays. Alignment of CCDs along their array or the alignment of arrays between each other are strictly checked in laboratories before the launch of the satellite. Viewing angles of each CCD are provided to the processing facilities. These angles are used in the viewing model, enabling facilities to compute the products that are distributed to users.

Because of launch conditions or simply because of aging, differences may be observed between current values of these CCD viewing angles and those measured before launch. Such variations are very complex to measure in order to correct the viewing model.

Figure 4.12 shows, for example, an alignment defect along pixels in a wide-field product. This defect has been detected analyzing the disparities with a reference image.

Figure 4.12. *Alignment defect along pixels of a wide-field segment*

Scanner instruments

Acquisition by a scanner instrument (for example, Landsat TM or ETM+) is performed by series of alternated scans (each one being called *swath*). To set successive scans as parallel as possible, the light path through the instrument is deviated by lateral correcting mirrors. To reconstruct the image, each swath is interpolated by polynomials adjusted on three dates that are measured when the oscillating mirror crosses three internal cells.

Errors in the measurement of these dates may lead to a shift of one or more swaths towards the successive ones (see the example in Figure 4.13).

Figure 4.13. *Swath shift in a Landsat TM image*
(Courtesy GAEL Consultant)

Other geometry errors have been observed in images acquired by a scanner instrument (see the non-exhaustive list given in Figure 4.14).

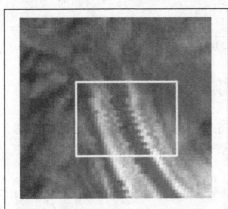

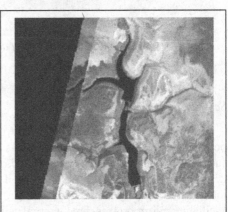

| a – Displacement along line. | b – Shift of channels. |

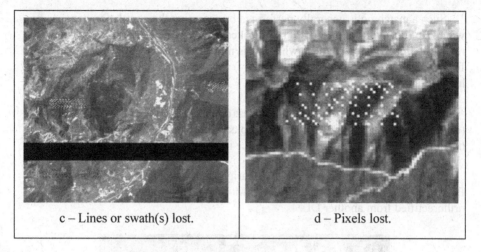

| c – Lines or swath(s) lost. | d – Pixels lost. |

Figure 4.14. *Some defects observed in Landsat TM images*
(Courtesy GAEL Consultant)

4.2.4. *Localization control and global models*

Absolute location checks (section 4.2.3.1), global defects (section 4.2.3.2) or defects with a magnitude at least greater than one pixel may be processed using an application that enables one to collect check points (section 4.2.2.4) and measuring the:

– quadratic mean of localization errors (equation 4.5);

– arithmetic mean of localization errors (equation 4.7).

To be significant, statistics will be based on at least 20 check points regularly distributed in the image and using reference cartographic documents of sufficient quality which have a well-known spatial accuracy.

4.2.5. *Internal geometry control*

Internal geometry checks will enable the detection of errors of sub-pixel size. This is, for example, required when assessing the CCD viewing angles that verify that viewing vectors intersect a plane tangent to the Earth according to a regular grid.

One of the most frequently used methods is the *disparity analysis*, in which we look for matching points (or a subset of regularly spaced points) of the image to be matched with their homologous points in a reference image.

Such a matching is generally performed by assessing the correlation between two windows (one in the image to be checked and the other one predicted in the reference image). One of these windows is moved across an exploration window to find the value of the displacement (dx,dy) for which the correlation index is the highest one.

Disparity analysis produces two images: one image of the vertical displacements (dx), and one image of the horizontal displacements (dy). Sometimes a third image may be computed, providing the confidence index of each matching.

Figure 4.15 shows results of a disparity analysis performed on one Landsat ETM+ scene orthorectified from one DEM with regard to the same Landsat scene orthorectified from another DEM.

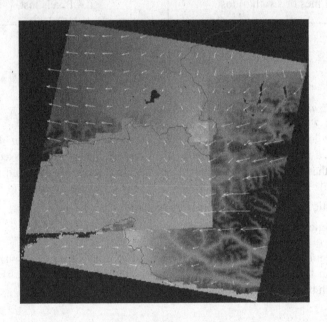

Figure 4.15. *Results of the disparity analysis on a Landsat ETM+ scene orthorectified with two different DEMs*
(Courtesy GAEL Consultant)

As may be observed, disparity analysis may produce very accurate results that enable the retrieval of the limits of the DEMs that have been used for the orthorectification, and also the relief (or more precisely the "anti-relief") in zones where only one DEM has been used.

4.3. Radiometry quality

4.3.1. *Radiometry quantities*

The spectral domain of concern is the solar spectrum. Therefore, we exclude the thermal domain for which the concept of apparent temperature is introduced. The energy flux ϕ corresponds to an energy that can be emitted, received or transported per unit of time and then expressed in Watts. The density of energy flux (or net flux) is then defined as the energy flux that goes through an elementary surface.

$$F = \frac{d\phi}{d\Sigma} \quad \left(W.m^{-2}\right) \tag{4.9}$$

The radiance corresponds to the energetic flux that goes through an elementary surface $d\Sigma$, following a direction θ to the normal to the surface and contained in a solid angle $d\Omega$. This radiance is expressed as:

$$L = \frac{d^2\phi}{d\Omega d\Sigma \cos\theta} \quad \left(W.m{-}2.sr{-}1\right). \tag{4.10}$$

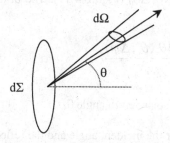

Figure 4.16. *Energetic flux through an elementary surface*

Also introduced now is the monochromatic radiance L_λ ($W.m^{-2}.sr^{-1}.\mu m^{-1}$):

$$L_\lambda = \frac{dL}{d\lambda} \tag{4.11}$$

For a satellite sensor, accounting for the spectral response R_λ of the instrument, it is useful to introduce the concept of equivalent monochromatic radiance L_{eq} ($W.m^{-2}.sr^{-1}.\mu m^{-1}$):

$$L_{eq} = \frac{\int L_\lambda R_\lambda d\lambda}{\int R_\lambda d\lambda} \tag{4.12}$$

Let us consider a parallel beam of irradiance Es, lighting the elementary surface under an incident angle θ_s.

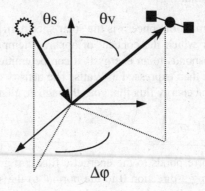

Figure 4.17. *The direct to direct path from the Sun to a satellite*

Let us consider the radiance $L(\theta_s,\theta_v,\Delta\varphi)$, reflected by the surface in a given direction, characterized by $(\theta_v, \Delta\varphi)$ (Figure 4.17). The bi-directional reflectance is defined as:

$$\rho(\theta_s,\theta_v,\Delta\varphi) = \frac{\pi L(\theta_s,\theta_v,\Delta\varphi)}{\mu_s E_s}$$

[4.13]

where μ_s is the cosine of the solar zenith angle θ_s.

If ρ is constant, whatever the incident angle and the reflection angle are, then the surface is Lambertian and the reflected radiance is isotropic. If the reflectance is non-Lambertian, a given target illuminated or viewed under different geometrical conditions does not give the same reflectance. One can then normalize the observed reflectance to the one we should observe with the Sun at its zenith and for a nadir view. This normalization requires a library of standard bi-directional models. The choice of the bi-directional model depends upon the type of surface. The surface classification can be conducted on color images using a spectral index which is quite independent of the geometry.

4.3.2. *Overview of the radiometric defects*

The various factors that affect the radiometric quality of the data, given some information about the applied corrections, are discussed below. The last example is more specific to aerial views.

4.3.2.1. *Diffraction and defocalization*

Using a schematic representation of the optics by a simple lens, an object point located at infinite only converges in the focal plane if the instrument is perfectly focused. On the other hand, because of diffraction, a point image corresponds actually to a small disk. These two defaults are generally well controlled.

4.3.2.2. *Polarization of the instrument*

The light is polarized; the intensity varies when rotating an analyzer in front of a camera viewing a natural scene. On the other hand, the optics of one instrument polarizes by itself. For the same incident light, the response of the instrument depends upon the polarization induced by its optics. Such a simple experiment enables the description of the intrinsic polarization of one instrument. Therefore, having some idea about the degree of polarization of natural scenes, we can apply a first order correction of additional energy due to the instrumental polarization.

4.3.2.3. *Stray light*

Within the optical assembly, undesirable reflections occur on the different inside walls (surfaces), as well as on the optical gun, even if all the surfaces are treated with anti-blooming material. For ideal optics, a well-collimated thin beam (as is obtained from a laser beam) only lights an elementary detector. In practice, one can observe a spatial distribution of light on the full image. This light distribution is characterized by the point spread function (PSF) which can be used to correct this stray light effect. This correction can be time-consuming, but it is a vital requirement for improving the radiometric quality.

4.3.2.4. *Aerial photos*

Aerial shots taken in conic geometry induce non-isotropic distribution of light densities in the image. Such an effect, called "vignetting" ("*vignettage*" in French), generally produces darker corners or sometimes a decrease in the brightness from the center to the image borders (see Figure 4.18).

Vignetting is due to the composition of atmosphere (humidity, aerosols, pollution, etc.), to the focal length used (a greater aperture at 153 mm produces more vignetting than a 210 mm focal), to the altitude of flight (below 500 meters, vignetting is less significant), to the location of the aircraft relative to the 0°C inversion layer, to the attitude of the aircraft, etc.

For better photographs, it is possible to use filters limiting the vignetting effect. Modern optic lenses are now available to compensate for this defect.

Figure 4.18. *Raw aerial photo (left) and after vignetting correction (right)*
(Courtesy TTI Production, France)

Many software tools enable correction of the vignetting defect. One of the most frequently used algorithms consists of computing light distribution statistics, both along lines and along columns, and interpolating a polynomial correction function that will register radiometry on the one observed at the centre of the image (see Figure 4.18).

4.3.3. *Calibration of the radiometric data*

4.3.3.1. *Radiometric calibration*

The energy (in Joules) received by a sensor is converted into digital counts (DC). Even without any incoming light, digital counts, DC_0, are recorded (dark current). This dark current is routinely measured by occulting the optics or, when it is not feasible, by measuring the dark signal (for example, during night over the oceans or over desert areas).

The calibration consists of associating DC with the desired radiometric value. To do so, the DCs are recorded in circumstances for which the incoming signal is predictable. Calibration can be achieved in the laboratory, in flight with an onboard calibration device, or over targets for which the optical characteristics are known (so-called vicarious calibration).

4.3.3.1.1. Laboratory calibration

Standard calibrated sources are used (standard lamp, power supplied by a well-controlled and stabilized current and, under theses conditions, stable during a given period of time), from which we can know accurately the monochromatic irradiance $(W.m^{-2}.\mu m^{-1})$ at a given distance (generally one meter). This irradiance is transformed into radiance. Two main experimental settings are used:

(i) A perfectly reflecting sphere is equipped with one or several standard lamps. Multiple reflections within the sphere produce a perfect isotropic radiance that is accurately known. The sensor that we want to calibrate views the interior of the sphere through a small hole. The bigger the sphere is, the better the accuracy will be because the relative impact of the entrance hole will be reduced.

(ii) The standard lamp lights a white standard panel which is perfectly reflecting and almost Lambertian (bi-directional properties of the panel reflectance are known for a second-order correction). The distance between the lamp and the panel is precisely known in order to perfectly calculate the incident irradiance on the panel.

The advantages of the laboratory calibration are: to offer well-controlled conditions, and to be repeatable. The accuracy decreases in the violet-blue spectral range with the sharp decrease of the lamp irradiances. If the calibration procedures are well conducted, one can expect to reach an absolute accuracy of 1% to 2%.

4.3.3.1.2. Onboard calibration

Laboratory calibrations are regularly used for aircraft instrumentation and can be performed on request. There are now onboard calibration devices for satellite sensors. The Sun is generally the direct source of light. Its light is reflected by a standard panel, well characterized in reflectance in order to know precisely the incoming radiance to the sensor.

4.3.3.1.3. Vicarious calibration

In absence of onboard calibration, or to double check it, vicarious calibration is used to predict the incoming light. This prediction is generally done during dedicated field campaigns during which all the components of the signal are well quantified: surface reflectance, influence of the atmosphere, etc.

4.3.3.2. *Spectral calibration*

Most of the observations are multi-spectral and it is necessary to characterize the spectral response of each band (at least, the mean wavelength). A laboratory monochromator can be used for pre-flight characterization. Increasingly, spectrometers or hyper-spectral instruments are used in space. For specific band settings, these instruments can be aimed at specific targets that offer high spectral variabilities with a known spectral response. It is primarily the case of the solar spectrum with the Fraunhoffer line as well as of well-defined atmospheric absorption lines (for example, the oxygen absorption around 761 nm). By matching these special spectral features, we can then achieve a vicarious spectral calibration.

4.3.4. *Atmospheric correction*

In a chapter devoted to the quality of the radiometric data, it is difficult to skip the problem of the atmospheric correction when the domain of interest concerns the Earth's surface. Atmospheric effects are major for oceanic observations: the atmospheric contribution represents between 60% and 80% of the signal. Land surfaces are more reflective. Nevertheless, the atmospheric correction is a key issue for the quality of the images that measure the Earth's surface radiances.

Let us use a simplified formulation of the signal to better understand the atmospheric correction problem:

$$\rho^* = T_g(\rho_{atm} + T_{atm}(\mu_s)\frac{\rho_G}{1-\rho_G S_{atm}}T_{atm}(\mu_v))$$ [4.14]

The gaseous transmittance T_g is associated with the ozone (Chapuis band between 0.55 μm and 0.7 μm) and is easy to correct. Most of the spectral bands avoid the gaseous absorption, and then $T_g = 1$.

ρ_{atm}, T_{atm} and S_{atm} are respectively the atmospheric reflectance (the satellite signal observed over a dark target), the atmospheric transmittance (the attenuation of the light on the atmospheric path), and the spherical albedo (the proportion of the light reflected by the surface and then backscattered downward). The atmosphere is composed of molecules for which the abundance (through the barometric pressure) and the scattering properties are well known (Rayleigh scattering). The molecular scattering is very effective in the blue band. Complementary to the molecules, the aerosols present a large spatio-temporal variability. Therefore, we need to characterize them. To achieve the remote sensing of aerosols, specific spectral bands are chosen, for which the surface contribution is minimum. They are located in the red and near-infrared for oceanic observations. Over land, vegetation is used for aerosol remote-sensing: vegetation is quite dark in the blue and red bands because of the photosynthesis. While the aerosols are characterized, atmospheric corrections schemes are applied.

The above atmospheric correction is relevant for a homogeneous Lambertian surface. The quality of the image is greatly improved and therefore atmospheric correction schemes are applied on an operational basis. For colored landscapes and high-resolution pictures (of the order of few meters), the atmospheric effects are more complex because of the so-called adjacency effect.

4.4. References

[CON 97] CONSEIL INTERNATIONAL DE LA LANGUE FRANÇAISE, *Terminologie de Télédétection et Photogrammétrie*, 1997, Paris, PUF.

[RIA 02] RIAZANOFF S., *SPOT 123-4-5 Geometry Handbook*, 2002, Toulouse, SPOT IMAGE.

Chapter 5

Understanding the Nature and Magnitude of Uncertainty in Geopolitical and Interpretive Choropleth Maps

5.1. Introduction

Since the early 1990s, spatial uncertainty has been considered a key part of spatial science. Efforts to quantify this uncertainty date to the early 1990s (for example, [CHR 82]). Since that time, a considerable amount of effort has been put forth to model spatial uncertainty, develop taxonomies, and describe its sources (for example, [LOW 99]). As a result of this effort, most practitioners understand that spatial uncertainty can affect the reliability of land management decisions made using digital spatial information.

Conventional texts on cartography (for example, [DEN 98]) identify two types of thematic maps – isarithmic and choropleth. Both types of maps indicate points that have the same attribute value for the theme of interest. Isarithmic maps are composed of lines – that is, isotherms – on which the value of the map attribute (which is always continuous) is constant along the same line, but potentially different for different lines (Figure 5.1). Choropleth maps are composed of polygons; the value of the attribute mapped – whether categorical or continuous – is considered uniform within a single polygon, but may be different from one polygon to another (Figure 5.1).

Chapter written by Kim LOWELL.

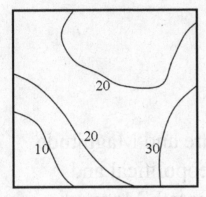 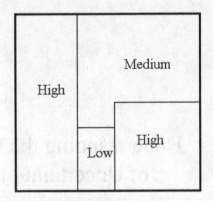

Figure 5.1. *Examples of isarithmic (left) and choropleth (right) maps*

Assessing uncertainty associated with different types of maps has been of considerable interest since the early 1990s. Indeed, much work on describing types and sources of uncertainty dates from that period – for example, [THA 92], [HUN 92]. However, it has rarely been recognized that the manner in which spatial uncertainty is quantified is very different for isarithmic and polygonal choropleth maps. This chapter will not address uncertainty related to isarithmic maps; a large body of scientific literature exists for this subject with particular emphasis on isarithmic topographic maps (see, for example, [CHI 99], [KRO 99], [KYR 01], [YAM 00] and their associated bibliographies). Instead, this chapter will address the nature and magnitude of errors on polygonal, choropleth maps. When estimating spatial uncertainty for such maps, polygonal choropleth maps must be subdivided into two subcategories that reflect different methods of construction.

The first subcategory of polygonal choropleth maps is herein termed "geopolitical" maps, which means that polygon boundaries are geopolitical and therefore exist by definition – for example, state, county, or township boundaries, and/or boundaries that separate the land owned by different individuals. The second subcategory of polygonal choropleth maps is herein termed "interpretive" maps because polygon boundaries have been identified semi-subjectively – for example, soil map boundaries, forest-type map boundaries. At first glance, both maps are similar and are of the form in Figure 5.1. They are composed of non-overlapping polygons, and each polygon is assigned an attribute value for the theme being considered – for example, population density, soil type. However, the manner in which uncertainty can be quantified is fundamentally different for these two subcategories of polygonal choropleth maps owing to the fact that for one type – the geopolitical map – one can be sure that a polygon boundary exists, but for the other – the interpretive map – one cannot assume that polygon boundaries truly exist in the real world.

5.2. Uncertainty in geopolitical maps

Geopolitical, polygonal maps are an integral part of conventional cartography. Most cartography texts use them as the geometrical base of choropleth maps – something that makes inherent sense. For example, if one is interested in producing a map showing the distribution of housing units for a particular state, it makes great sense to show the number of housing units for well-defined, well-accepted sub-units of the state – counties, for example. This is because the boundaries of each spatial unit are well defined and likely to be clearly identified on the ground, and one is also assured that each house belongs to one county only (except in very anomalous conditions). Moreover, a considerable amount of information, such as census and landholding information, is collected using the same spatial units, so there is an abundance of information available for these spatial units.

Relevant to this chapter, the use of geopolitical maps as the geometric base maps of thematic choropleth maps considerably simplifies the characterization of uncertainty. When one uses such maps, uncertainty can be divided into two separate themes, both of which are relatively simple to characterize due to the fact that one is using a base map whose internal polygon boundaries are known to exist. This allows one to consider *locational* uncertainty and *attribute* uncertainty separately, knowing that the quantification of each is relatively straightforward.

5.2.1. Locational uncertainty in geopolitical maps

In examining a choropleth map based on the spatial elements of a geopolitical map, one need only question whether or not a boundary is located in the correct position – one does not need to be concerned with the question "Does this polygon boundary and the polygon itself truly exist?"[1] Consequently, locational uncertainty is related primarily to well-defined factors: map projections, map scale, data acquisition, and data processing. Each of these is addressed in turn.

Map projections affect locational accuracy because of the distortion inherent in representing a three-dimensional object – the earth – on a two-dimensional medium, such as a piece of paper. The characteristics of specific projections and the location of the area of interest will determine the effect of map projections on locational uncertainty.[2] For example, using a cylindrical projection that has the equator as a

---

1 This statement intentionally ignores map-updating. If an area has been subdivided or agglomerated due to a new political territory division or the sale/acquisition of property, then it is possible that an existing map contains boundaries that do not necessarily exist. However, out-of-date maps are not considered herein since updating cycles tend to be controlled by non-cartographic considerations and are not a problem in cartographic fundamentals.

2 For a good reference on map projections, see [SNY 87].

reference parallel to map areas close to the equator will result in the map having virtually no distortion in the east-west direction and very little distortion in the north-south direction. However, the same projection used to map areas above 45° N or below 45° S would result in maps having potentially excessive amounts of east-west distortion (or uncertainty), and even higher amounts of north-south distortion/uncertainty. Moreover, the fact that the earth is not a perfect sphere or ellipsoid will compound whatever uncertainty is inherently related to the map projection employed to produce a map. Note that relative to map projections and other causes of locational uncertainty for geopolitical maps, "uncertainty" is reasonably synonymous with "inaccuracy", although this is not always the case. This will be discussed further subsequently.

Map scale effects locational uncertainty of geopolitical maps in a number of ways. The most obvious is the geometric simplification that occurs when one changes from a large-scale map – for example, 1:1,000 – to a small-scale map – for example, 1:100,000. When converting a large-scale map to a smaller scale, it becomes impossible to retain all the details of all the features mapped, particularly for crenulated features such as rivers.[3] Moreover, the physical width of a line on the map itself becomes an issue. For example, 1-mm-wide lines on a 1:1,000 map represent 1 m on the ground whereas the same line represents 100 m on the 1:100,000 map. Converting a map from a large to a small scale, while retaining the same line width on a map, necessarily adds locational uncertainty to the latter.

As for data acquisition and locational uncertainty, much depends on the instrumentation used to capture the data. For example, a feature whose position is recorded using a global positioning system (GPS) receiver without differential correction will have a greater amount of locational uncertainty/inaccuracy than if the same information is subject to differential correction. Similarly, the accuracy and precision of the distance between two objects as determined using laser positioning would be superior to that obtained by using a tape measure.

Locational uncertainty on geopolitical maps is also affected by "data processing" – herein defined as the conversion of physical maps into digital information. Locational uncertainty related to data processing is affected in part by map projections, map scale, and data acquisition methods. However, it is probably more affected by the equipment used to convert cartographic information to a digital format, and the skill of the operators who use that equipment. For example, a digitizing tablet operator who does not conscientiously digitize enough points to represent a highly crenulated line will have added locational uncertainty to the map. However, even an extremely conscientious operator will not be able to locate a line

---

3 For more information on the effects of changing scale, readers are referred to literature on map generalization – e.g., [MAR 89].

"exactly correctly" if the width of the line on the map exceeds that width of the cross-hairs on the cursor of the digitizing tablet. A number of studies on data processing and uncertainty have been conducted – for example, Caspary and Scheuring [CAS 93].

Finally, locational uncertainty can also be affected by the nature of the boundaries themselves; this is the situation mentioned earlier in which uncertainty and inaccuracy are not synonymous. Although geopolitical boundaries are known to exist, there are special cases in which the boundaries can change radically or are simply difficult to locate in the "real world". For example, some legal boundaries are defined as being the center point of a river – the boundary between the states of Vermont and New Hampshire, for example, is delimited by the White River. Should the location of the river change due to extreme flooding events or simple erosion, the location and form of a boundary could change. Of greater concern, however, is that though a boundary defined as "the center point of a river" is straightforward, locating such a point on the ground may not be, particularly if a river course is excessively crenulated. Thus, for such boundaries, locational uncertainty is not the same as inaccuracy because the centre point of a river is not an indisputable location.

5.2.2. *Attribute uncertainty in geopolitical maps*

Quantifying attribute uncertainty for geopolitical maps is primarily a statistical issue. In knowing that the boundaries of the spatial units of interest exist and have real locations, one need "merely" obtain information about the theme of interest for each spatial unit.

If one is dealing with categorical information – for example, nominal or ordinal attributes – there is virtually no associated uncertainty since something either has or does not have the characteristic of interest. For example, if one wishes to map all counties through which a major highway passes, one need only record for each county whether or not such a condition exists. Similarly, producing maps of areas having a particular type of zoning (or combination of zoning) is straightforward.

If one is working with continuous – that is, interval/ratio – attributes, attribute uncertainty is only slightly more complicated. For example, if one wishes to map the average house price by county in a particular state, one must obtain information about the average house price for each county. Obtaining such information is relatively straightforward and a number of ways of obtaining such information exist. For example, most county or municipal governments keep records of property transfers. Similarly, real estate corporations often keep such records in order to track market trends.

What is important to understand relative to attribute uncertainty, however, is that virtually all such information represents only a sample of the information desired rather than a complete census. Property transactions for a single year, for example, do not indicate the average sale price of all property in a county since not all properties will have been sold in a given year. This means that the estimate assigned to a given spatial unit has some level of confidence associated with it. Fortunately, such uncertainty can be described using standard statistical tools that are well adapted to describing sampling uncertainty and population variability – for example, coefficient of variation, standard error.[4] Of note in employing such tools is that the attribute uncertainty may not be the same for all spatial units. If the theme of interest within a particular spatial unit has more variability in the attribute studied than in another attribute, uncertainty will be higher. Similarly, a lower sampling intensity will result in an increased amount of attribute uncertainty for a given spatial unit.

5.3. Uncertainty in interpretive maps

At first glance, interpretive polygonal maps are indistinguishable from geopolitical polygonal maps. Both are composed of spatial units – polygons – that are labeled with a particular attribute. However, the differences in the way that they are constructed leads to radical differences in the magnitude and type of spatial uncertainty associated with each, and the way that it must be estimated.

5.3.1. *Construction of interpretive polygonal maps*

Whereas geopolitical thematic maps employ spatial units whose boundaries exist by definition, interpretive maps are composed of spatial units whose boundaries only exist relative to some classification system.[5] Soil maps are a typical example. Within the subject of pedology, soil taxonomy occupies a prominent position. In the creation of soils maps, therefore, it would at first appear to be a relatively simple matter to produce a map based on a particular soil taxonomy. That this is not the case, however, is related to two factors.

First, uncertainty in the map production process is related to the taxonomic system itself, as well as to the minimum mapping unit (MMU)[6] employed to produce the map. At one extreme, if one is mapping the soils of a 1,000 ha parcel

---

4 For more detailed information, readers are referred to any introductory text on descriptive statistics.
5 This has long been recognized in cartography in general – see, e.g., [JOH 68].
6 For a better understanding on how the MMU unit can affect various spatial analyses, readers are referred to literature on the modifiable aerial unit problem (or MAUP) – e.g., [CLA 76], [OPE 84], [HOL 96].

with an MMU of 250 ha, at most the soils map will contain four polygons *regardless of the detail of individual soil descriptions* – that is, the spatial resolution is 250 ha. Similarly, if the soil taxonomy is as simple as a three-class system – soil, rock, and water – then the resulting map will only have three different types of polygons – that is, the attribute resolution is three classes – although many such polygons may be present, depending on the MMU. These factors are a sharp contrast relative to geopolitical maps for which it is known *a priori* how many polygons will be present on the final map and where their boundaries will be located. With interpretive maps, the number of polygons, as well as their individual form, will vary with the classification system and the MMU employed. Hence the concept of "locational uncertainty" is of limited use with interpretive maps. Instead, "existential uncertainty" is of more interest because a particular boundary exists only relative to a given classification system and MMU.

Secondly, to understand the uncertainty associated with interpretive maps, one must examine the methods by which they are constructed. Generally, interpretive maps are produced using some sort of remotely sensed images – for example, aerial photographs, satellite images. Such images are classified relative to the classification system and MMU of interest. This classification is generally done subjectively by human beings who have specific training, and the data used in the classification process are supplemented by a limited amount of ground-based data collected for this purpose. This factor combines with those described in the preceding paragraph to produce maps that have hidden, potentially large, sources of uncertainty.

To illustrate this, consider the soil map example. Because a soil taxonomy is developed independently from a MMU, the taxonomy often has an attribute resolution that requires a finer spatial resolution than a map can support. Consequently, "cartographic soil taxonomy" may be developed by combining soil types such that a "cartographic soil type C" may be something like "at least 80% Type A with no more than 20% Type B". Hence a mapped polygon of Soil Type C may in reality – that is, on the ground – be composed of two distinctly different soil types. This causes problems particularly during the human interpretation of the remotely-sensed images because human beings have a tendency to organize spatial objects in a way that "makes sense" to them. For example, Figure 5.2 shows a series of black and white patches. However, because of human tendencies to organize space, most people will "clearly see" a Dalmatian dog, even though the parts of the dog – for example, the head, the legs – are not explicitly drawn. In developing polygon boundaries for interpretive maps where basic taxonomic units must be combined due to the MMU, different people will not organize space in the same way. As an example of the effect of this, Figure 5.3 shows a spatial arrangement of triangles and squares and three equally valid ways of dividing the space to produce one polygon that is "predominantly squares" and one that is "predominantly triangles" while respecting the constraint of a MMU that occupies no less than one-fifth of the total space.

Figure 5.2. *Black and white patches that resemble a Dalmatian dog*

5.3.2. Uncertainty in boundaries of interpretive polygonal maps

The factors discussed in the previous section combine with each other to change the way that one must think about "locational uncertainty" on interpretive choropleth maps. For geopolitical maps, map boundaries have real-world counterparts regardless of the MMU employed (although the selected MMU and map scale of a geopolitical map may not allow the display of all boundaries). Moreover, given enough resources, boundaries on geopolitical maps can be located on the ground. Conversely, boundaries on interpretive maps only exist relative to a given classification system and MMU, and are dependent on how a given cartographer decides to organize individual spatial elements into polygons (see Figures 5.2 and 5.3).

What makes this particularly difficult when dealing with uncertainty on interpretive maps is that each boundary may or may not have a real-world counterpart, and the ease of locating in the real world those boundaries that do exist varies. For example, the real-world boundary between rock and water may be quite evident, but the real-world boundary between Soil Type A and Soil Type C (still defined as at least 80% Type A with up to 20% inclusions of Type B) may be impossible to find. Such a boundary might even be considered to be merely an artifact of the cartographic construction process! Thus, in the case of interpretive polygonal maps, "locational uncertainty" gives way to "existential uncertainty" by which the likelihood that a given boundary exists varies with the characteristics of the phenomenon being mapped and associated cartographic parameters.

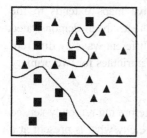 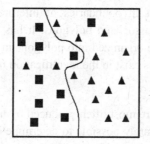 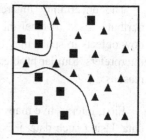

Figure 5.3. *Three different, equally valid interpretations of the same objects*

Nonetheless, even with interpretive cartographic maps, one can have high confidence that certain boundaries really do exist in the "real world". For such boundaries, the concept of "locational uncertainty" is certainly present, but its causes are very different from those that cause locational uncertainty on geopolitical maps. For example, if one travels by boat across a clear lake, to an area of swamp, to an island that is definitely dry land, three different types are definitely present: lake, swamp, land. By definition, therefore, there must be two boundaries present. However, the real-world swamp-island boundary is likely to be easier to find than the lake-swamp boundary. In both cases, however, it will be impossible to locate exactly the boundary between the two types on either side of the boundary. Thus one must accept that with interpretive maps, the real-world counterpart of a map boundary usually is not a widthless line as is the case for geopolitical maps. Instead, boundaries on interpretive maps are transition zones whose width varies depending on the nature of the classification system employed to produce the map, and the characteristics of the phenomenon being studied in the area being mapped.

The discussion in this section has suggested that the uncertainty associated with boundaries on interpretive maps is due only to cartographic parameters on which map construction is based and the physical characteristics of the area being mapped. In addition, however, a limited amount of boundary uncertainty on interpretive maps will be related to the same factors mentioned relative to geopolitical maps – that is, data processing, cartographic projections, etc. However, the magnitude of the uncertainty caused by such factors on interpretive maps is trivial in comparison to the uncertainty caused by the other factors mentioned in this section.

5.3.3. *Uncertainty in attributes of interpretive polygonal maps*

The uncertainty inherent in boundaries of interpretive maps extends to the attributes used to label the associated polygons. It has already been demonstrated how objects in space may be organized into polygons in different ways by different interpreters. Similar problems exist in the identification of attributes for two primary reasons.

First, interpretive maps are most often produced by interpreting remotely-sensed data. Using such data, it is rarely possible to determine exactly what is present at a given location. For example, suppose that one wants to produce a forest map that shows polygons of pure spruce (PS), pure fir (PF), and spruce-fir mix (SFM). Even if two different interpreters are in perfect agreement as to where the boundaries between two types should be placed, one might label a given polygon PS whereas the other may believe that some fir is present and label the polygon SFM. Hence there might not be perfect attribute agreement – note that the words "attribute *error*" were not used as there is no "truth" against which a polygon label can be compared – because the objects examined to produce the map look similar on the medium being used to produce the map. Attribute uncertainty also occurs if the two interpreters agree on the attributes of two polygons, but do not agree on where the boundary between the two should be placed. Figure 5.4 shows a situation in which two interpreters agree on the attributes for two polygons, but disagree on boundary placement, which leads to attribute uncertainty where the polygons of the two interpretations overlap. This is most likely indicative of the real-world boundary between the two types being a transition zone rather than a widthless line as discussed in the previous section – a possibility that has long been recognized for a number of years (for example, [LEU 87]).

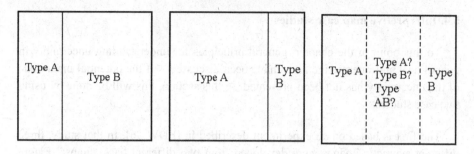

Figure 5.4. *Perfect attribute agreement, but lack of geometric agreement leading to attribute disagreement (left: one interpretation; center: another interpretation; right: result of both interpretations)*

The second cause of attribute uncertainty on interpretive maps is related to the characteristics of an area being mapped relative to the class boundaries of the classification system employed. For example, "forest" might be defined as being "20% tree cover with a mean tree height of at least 2 m". While such a definition is very clear, in reality determining if an area is only 19% tree cover or if it achieves the necessary 20% threshold depends on the skill of the interpreter and the manner in which the interpreter mentally organizes the trees in the space. Similarly, it is difficult to determine if the mean tree height of a given area is exactly 2 m – particularly when the heights of trees in a particular area vary widely. The MMU has a pronounced effect here as the area over which a photo-interpreter will estimate the average tree height will change with the MMU.

Seemingly, these problems of attribute uncertainty could be resolved if one had enough resources to visit all points on the ground. After all, this would allow one to determine, for example, whether or not a given area had sufficient tree density or height to be classified as forest. Such an idea is incorrect, however. In visiting a given point on the ground, one loses all ability to respect the MMU that is essential for map construction. Moreover, one may have difficulty controlling the scale of observation in the field. For example, if one is going to measure topographic slope on the ground, one's observations will be based on a much larger distance if one is standing in an open field than if one is standing in an extremely dense forest. What this means is that information taken on the ground is only reliable for one scale and paradoxically might be of poorer quality for certain applications than data mapped at a coarser scale.

5.4. Interpretive map case studies

To this point in the chapter, general principles for understanding uncertainty in geopolitical and interpretive maps have been discussed, but the potential magnitude of the uncertainty has not been presented. In this section, this will be done by using two case studies.

The first is based on an experiment described in [EDW 96]. In that study, three different synthetic images were developed from two different base "maps;" Figure 5.5a shows one of the base maps developed that contained six different classes. "Textures" were applied to each class on the base maps to produce images that resemble aerial photographs (Figure 5.5b). Subsequently, nine different individuals were asked to identify homogeneous areas on the image and were given the information that there were between three and ten classes present, and that the MMU was 1 cm on the computer screen used to project the image; Figures 5.5c and 5.5d show two of the nine interpretations/maps. Interpreters, in general, correctly identified geometric characteristics in the "northern" part of the image – for example, the big oblong polygon in the "northwestern"-central portion of the image, the dark island in the "northeastern" portion of the image. However, certain features were not correctly identified in the "northern" portion of the image. For example, the small triangular polygon in the "northwestern" corner was not identified by all interpreters, and very few interpreters identified the polygon in the centre of the base map along the "northern" edge of the base map. In general, however, for the "northern" section of the image, interpretations/maps matched the base map (which was considered to be "ground truth") reasonably well, and interpretations/maps showed reasonable consistency from one to another. In the southern section of the image, however, interpretations/maps varied widely from each other, as well as from the mask. In particular, the "southeastern" section was extremely problematic for the interpreters. Notably, for that section interpreters were reasonably consistent with each other, but not with the base map (variants of Figure 5.5c were the most common interpretation for that area). Similarly, the large, dark polygon along the "south central" boundary of the base map was generally combined with an adjacent polygon, and the attribute mislabeled. As a measure of the magnitude of the global uncertainty, the average agreement of the photo-interpreters with the base map was about 50%. Thus, on any single interpretation/map, the confidence that a map consumer could have in the mapped class actually being present at a given point is only 50%!

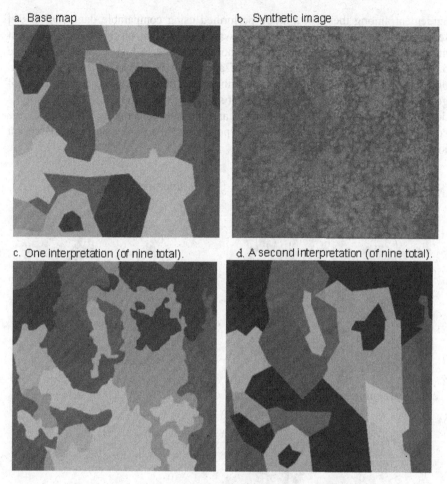

Figure 5.5. *Data from an experiment by Edwards and Lowell [EDW 96]*

The second case study is based on unpublished data compiled by the author of this chapter. Three forest maps of the same area were produced by three different interpreters using identical cartographic techniques and aerial photographs (Figure 5.6). Hence, whereas Figure 5.5 presents synthetic data, Figure 5.6 presents real forest maps produced by professional photo-interpreters. Among the three interpretations/maps, it is notable that there is relatively little agreement, either geometrically or for polygon attributes. Only very dark polygons are clearly in agreement; these happen to represent lakes and are the easiest features to identify in this particular area. It is also interesting that the "style" of interpretations/maps appears to be very different – Interpreter 3 identified far fewer polygons than did the other two interpreters. Although these maps were not overlaid in order to determine agreement among them, in a study performed using comparable data, [DE G 99] found that only 15% of a forested area was in agreement on three different forest maps produced in the same manner as those presented in Figure 5.6. From Figure 5.6 it is apparent that the low level of agreement is due at times to locational uncertainty of lines, at other times to attribute uncertainty due to different labels of geometrically equivalent polygons, and at still other times the overall uncertainty is due to an interaction of both locational and attribute uncertainty. Regardless, the confidence one can have that the forest type mapped at a given point is truly present in the real world is extremely low.

a. Interpretation 1.   b. Interpretation 2.

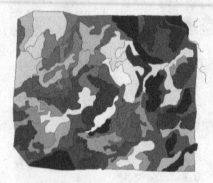 

c. Interpretation 3.

Figure 5.6. *Three different interpretations/maps of the same aerial photographs produced by professional photo-interpreters. For orientation, note the black polygons in the northeastern and east-central portion of the interpretations*

These case studies have demonstrated that the uncertainty of interpretive polygonal choropleth maps can be extremely high. However, even high uncertainty does not mean that the maps are of no use. Moreover, the level of uncertainty discussed here will not necessarily be present on all interpretive maps. For example, [LEG 96] noted general agreement among landscape units as identified from aerial photographs; attribute uncertainty was not examined.

Regardless of the level of uncertainty potentially present on a map, whether or not a map is of use depends on the answer to three related questions. First, is there a better source of information? Despite problems highlighted previously, this author has not found a better source of spatial information for large areas than the forest map. Hence, despite high levels of uncertainty, the forest map remains the best source of information for forest planning and a host of other activities. Secondly, what is the application for which the map is being used? Polygons on the forest maps employed as examples herein are defined based on the species mix, height, density, and age of the tallest trees in the polygon. Whereas agreement for all four of these attributes is relatively low at 15%, if one only wants to know where there is a high concentration of fir, for example, then the level of certainty is likely to be much higher. Thirdly, at what scale does information have to be reliable to be of use? Experience has shown that though agreement among photo-interpretations is low for individual points on forest maps, over a large enough area uncertainty is much lower. Thus, if one is doing forest management planning for a 10,000 ha parcel, for example, the forest maps used as examples herein are appropriate for this purpose. However, if one is using such maps to identify wood volume for individual 10 ha blocks, these same maps are not appropriate for that purpose.

5.5. Conclusion

This chapter has explained how the determination of the magnitude of uncertainty on a thematic polygon map depends very heavily on whether the map in question is a geopolitical map whose boundaries exist by definition, or if it is an interpretive map for which polygon boundaries will also have existential uncertainty. If the map is geopolitical, then quantification of uncertainty is relatively straightforward – geometric uncertainty is the difference of the true location of each line with its mapped position; categorical attributes will have little, if any, associated uncertainty, and quantitative attribute uncertainty is dependent on sampling considerations that can be described statistically. Conversely, interpretive maps will have considerably more inherent uncertainty that is much more difficult to quantify. Nonetheless, two case studies suggest that the potential uncertainty on interpretive maps – defined as disagreement among interpretations – is potentially huge. Adding to this problem is that taking extensive ground-based observations – that is, "ground-truth" – does not provide a means to determine where interpretive maps are "right" or "wrong" because the scale of observation of the ground-based information is unlikely to be at an appropriate scale for the map under study.

5.6. References

[CAS 93] CASPARY W. and SCHEURING R., "Positional accuracy in spatial databases", *Computers, Environment and Urban Systems*, 1993, vol 17, p 103–10.

[CHI 99] CHILÈS J.-P. and DELFINER P., *Geostatistics: Modeling Spatial Uncertainty*, Series in Probability and Statistics, 1999, New York, John Wiley & Sons.

[CHR 82] CHRISMAN N., Methods of Spatial Analysis Based on Error in Categorical Maps, 1982, PhD thesis, University of Bristol, England, p 26.

[CLA 76] CLARK W.A.V. and AVERY K.L., "The effects of data aggregation in statistical analysis", *Geographical Analysis*, 1976, vol 8, p 428–38.

[DE G 99] DE GROEVE T., L'incertitude spatiale dans la cartographie forestière, 1999, PhD Thesis, Laval University, Québec, Canada.

[DEN 98] DENT B.D., *Cartography: Thematic Map Design*, 5th edn, 1998, Reading, Massachusetts WCB/McGraw-Hill.

[EDW 96] EDWARDS G. and LOWELL K., "Modelling uncertainty in photointerpreted boundaries", *Photogrammetric Engineering and Remote Sensing*, 1996, vol 62, number 4, p 377–91.

[HOL 96] HOLT D., STEEL D.G., TRANMER M. and WRIGLEY N., "Aggregation and ecological effects in geographically based data", *Geographical Analysis*, 1996, vol 28, number 3, p 244–61.

[HUN 92] HUNTER G.J. and BEARD M.K., "Understanding error in spatial databases", *The Australian Surveyor*, 1992, vol 37, number 2, p 108–19.

[JOH 68] JOHNSTON R., "Choice in classification: the subjectivity of objective methods", *Annals of the Association of American Geographers*, 1968, vol 58, p 575–89.

[KRO 99] KROS J., PEBSEMA E., REINDS G. and FINKE P., "Uncertainty assessment in modelling soil acidification at the European scale: a case study", *Journal of Environmental Quality*, 1999, vol 28, number 2, p 366–77.

[KYR 01] KYRIAKIDIS P.C. and DUNGAN J.L., "A geostatistical approach for mapping thematic classification accuracy and evaluating the impact of inaccurate spatial data on ecological model predictions", *Environmental and Ecological Statistics*, 2001, vol 8, number 4, p 311–30.

[LEG 96] LEGROS J.-P., KOLBL O. and FALIPOU P., "Délimitation d'unités de paysage sur des photographies aériennes: éléments de réflexion pour la définition d'une méthode de tracé", *Étude et Gestion des Sols*, 1996, vol 3, number 2, p 113–24.

[LEU 87] LEUNG Y., "On the imprecision of boundaries", *Geographical Analysis*, 1987, vol 19, p 125–51.

[LOW 99] LOWELL K., "Effects of adjacent stand characteristics and boundary distance on density and volume of mapped land units in the boreal forest", *Plant Ecology*, 1999, vol 143, p 99–106.

[MAR 89] MARK D.M. and CSILLAG F., The nature of boundaries on 'area-class' maps, *Cartographica*, 1989, vol 26, p 65–78.

[OPE 84] OPENSHAW S., *The Modifiable Areal Unit Problem*, Concepts and Techniques in Modern Geography (CATMOG) 38, 1984, Geo Books, Norwich, UK.

[SNY 87] SNYDER J.P., Map Projections – A Working Manual, United States Geological Survey Professional Paper 1395, 1987, Washington. United States Government Printing Office.

[THA 92] THAPA K. and BOSSLER J., "Accuracy of spatial data used in geographic information systems", *Photogrammetric Engineering and Remote Sensing*, 1992, vol 58, number 6, p 835–41.

[YAM 00] YAMAMOTO J.K., "An alternative measure of the reliability of ordinary kriging estimates", *Mathematical Geology*, 2000, vol 32, p 489–509.

Chapter 6

The Impact of Positional Accuracy on the Computation of Cost Functions

6.1. Introduction

As long as statistical methods are applied, there is concern about data quality. Although data quality is an old scientific subject, it has recently received much more attention [GOO 98; SHI 03; FRA 04]. The main driving force is that there is an increasing use of data, and that, in particular, sharing of data has grown. This, obviously, limits liability. Also, the scientific field focusing on data quality has grown. Frank [FRA 98] notes that data quality is often described in terms of lineage, that data quality descriptions are producer-oriented, that data quality descriptions are not operational, and, sometimes, that data quality descriptions are not quantitative. He advocates the meta-model of data quality. This includes concepts such as the world (W), observations O1, O2, etc., Data D1, D2, etc., function of interest f, and errors e1, e2, etc. This chapter will proceed by distinguishing between different types of error (that is, bias versus random variation) and dealing with the propagation of error.

Elements of data quality typically include the positional accuracy for well-defined points, including polygons, the quality of transfer functions and other models, the linking of data quality of inputs to data quality of the results (error propagation), the building of formal models for relationships between reality, geographic data and use, definition of user's query, and definition and evaluation of the error function.

Chapter written by Alfred STEIN and Pepijn VAN OORT.

We will follow the statement that data quality descriptions must be made to suit the intended use of the data. Recent developments concern the use of ontologies in the space-time domain [VAN pre], the development of fuzzy and vague methods [DIL sub], and the formulation of a variance-covariance equation for polygons in a geographical information system (GIS) [VAN 05]. In this chapter we present a special example of the variance-covariance equation, assuming equal variances in all directions. We apply this to a study on soil sanitation in the Netherlands.

Soil contamination causes general problems in society. Among other problems, contaminated soil may have an effect on the health of those living near the location where the soil is contaminated, the quality of the crop, and the quality of the food. In soil sanitation, polygons are often used to delineate sub-areas in a contaminated site that require special treatment. Modern engineering approaches include the use of cost models to calculate the total costs of remediation of the soil.

6.2. Spatial data quality

Spatial data are different from ordinary data in the sense that the location where the data are collected is an integral part of the data. Therefore, spatial data are characterized by their value and by their location. The value can be a quantitative value, such as the PM10 content, the soil pH, groundwater depth, or degree of pollution, or a qualitative label, such as land use, land cover, geological unit, the type of building or the catchment area. The location of spatial data is usually expressed in terms of coordinates with respect to an arbitrary origin. Coordinates can be expressed as a one-dimensional coordinate, such as the distance along a transect, in two dimensions, such as the location in a plane, or in three dimensions, such as the location in a three-dimensional volume. Modern approaches now also allow the inclusion of issues of time, leading sometimes to a four-dimensional coordinate vector.

Any use of spatial data is affected by the quality of the data. The quality may then refer to the quality of the value (how precise the value is, how accurate the label is), as well as to the precision of the coordinates. These issues are generally addressed as issues of spatial data quality. As a quantitative expression for quality may not be sufficient, the causes of lack of quality are important sources of information; spatial data quality distinguishes seven major points of focus. Positional accuracy addresses the precision and accuracy of coordinates, whereas attribute accuracy concerns the precision of the attributes. In addition, temporal accuracy addresses issues of change in the data, and logical consistency concerns the question of whether collected data are related to other data in a logical sense. Lineage concerns the question of how the data are collected and the method of how the data have been entered in a computer program. Semantic accuracy addresses the question of whether the data really express what one may have in mind, and semantic completeness addresses the question of whether there is anything more to add to the data. In this chapter, attention will be paid to positional accuracy, as well as to its consequences on calculations with a cost model.

6.2.1. *Positional accuracy*

Positional accuracy is the accuracy in the coordinates that describe an object. In this chapter we focus entirely on data in two dimensions, that is, on data with two coordinates. The procedures described can be easily extended to more (or fewer) coordinates. We distinguish between point objects, line objects, and area objects. For a point object, let (x,y) be the coordinates of the object in the real world. Then the coordinates in a system are described with $(x + \sigma_x, y + \sigma_y)$ (Figure 6.1a). The values of σ_x and σ_y may be either known from previous studies, or from specifications of, for example, a global positioning system (GPS), or they may be derived from the collected data in relation to reference measurement. The positional accuracy is then $PA_P = \sqrt{(\sigma_x^2 + \sigma_y^2)}$. A line object (Figure 6.1b) supposedly consists of n point objects, (x_i, y_i), all with corresponding accuracies σ_x and σ_y. The positional accuracy is then $PA_L = \sqrt{n(\sigma_x^2 + \sigma_y^2)}$. Finally, an area object (Figure 6.1c) consists of a closed line. As a line, it has as a positional accuracy $PA_A = \sqrt{n(\sigma_x^2 + \sigma_y^2)}$. Other models have been proposed, for example, Hausdorff distance-based in BelHadjAli [BEL 02].

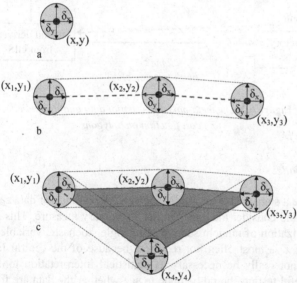

Figure 6.1. *Positional accuracy for point, line and area objects*

6.2.2. *The meta-model for spatial data quality*

The meta-model for spatial data quality is in fact an abstract, formalized model of the real world. It contains a model of the data, models of observation, and a correspondence process that links the world with the data and the models of the function of interest of a potential user. We will illustrate the meta-model by a simple model (Figure 6.2), where a world object (unambiguously agreed by experts), for example, a well-defined polygon, is represented by its truth (accepted as satisfactory by experts), namely its set of coordinates in a two-dimensional space. We also suppose that there is some function of interest; in this study this will be the cost function for soil sanitation. If the polygon is to be used, one may prefer to use a GIS, where this object is stored, instead of going to the field again, and to check for the function in the GIS. We would want to have not just an answer to the question, but also an indication of the quality of the answer.

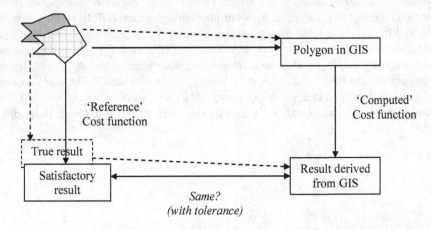

Figure 6.2. *The meta-model for spatial data quality applied to a cost function on polygons*

6.2.3. *Error model*

To describe a statistical error model, we consider a range of data z_i of n repeated observations, and consider the mean, m_Z as a summary measure. This is valid if the data are a realization of random sample of a single stochastic variable, Z. An exact distribution of Z is most often not required, because of the central limit theorem. Also, it may not really be necessary as statistical interpretation tools are robust. However, careful testing should be done to see whether the data are from a random sample. Several issues can be addressed by using boxplots, such as the presence of outliers, or the occurrence of systematic change.

To set up a full error mode, we suppose that there is a true value ζ and that X is a measurement variable for this value. This variable should to have a finite expectation, the so-called measurement expectation: $EX = \mu$. The measurement variance is then given by $E(X - \mu)^2 = \sigma^2$. We distinguish between the *systematic* error (the bias), being the difference between μ and ζ: $\delta = \mu - \zeta$, and the *random* error: $e = X - \mu$. Combining this we see that the measurement variable is the sum of true variable, the bias, and the random variable:

$$X = \zeta + \delta + e \qquad\qquad [6.1]$$

We notice that the common assumptions apply, namely that $E[e] = 0$ and $\text{var}[e] = \sigma^2$.

It depends on the circumstances whether δ should be taken into account; if $\delta < \frac{1}{2}\, \sigma/\sqrt{n}$, then the random errors dominate the mean, and δ can be ignored. Also, if we consider differences only between two variables with the same δ, then δ can be ignored as it disappears from the differencing. In many studies, systematic and random errors can not be completely separated.

Measuring the quality is done by assessing the discrepancy between two datasets in various ways, depending upon available world data. Typically, use is made of an independent test set, an independently selected sub-set, or a leave-one-out procedure. If it is done this way, then we compare observed (x_i) data with real world (t_i) data, for example, using one of the following measures:

ME: mean error: $(1/n)\, \Sigma_i (x_i - t_i)$ $\qquad\qquad [6.2a]$

MSE: mean squared error: $(1/n)\, \Sigma_i (x_i - t_i)^2$ $\qquad\qquad [6.2b]$

MAE: mean absolute error: $(1/n)\, \Sigma_i |x_i - t_i|$ $\qquad\qquad [6.2c]$

If a large bias exists, then the ME is large with respect to the actual measurements, whereas if a large uncertainty exists, then both the MSE and MAE are large.

6.2.4. Error propagation

Next, one may question how important the uncertainty in the data is. Small errors may be negligible in subsequent calculations, whereas large errors may have a significant impact. This question is answered by means of an error propagation ([HEU 98]). To answer the question, we assume that a given function f exists and that in this modeling it is considered to be a continuously differentiable function of n observable variables: $\zeta_1, \zeta_2, ..., \zeta_n$. The function is then given by $z = f(\zeta_1, \zeta_2, ..., \zeta_n)$.

Also, there are n observables $X_i = \zeta_i + \delta_i + e_i$ with $E(e_i) = 0$ and $var(e_i) = \sigma^2$. As the aim is to estimate z, we need an estimator Z^*, as well as the variance of Z^*. As a first guess for Z^*, we take $Z^* = f(X_1, X_2, ..., X_n)$; we therefore measure $X_1, ..., X_n$ once, and determine Z with f. However, as we may not have taken the measurements in the point $(\zeta_1, \zeta_2 ..., \zeta_n) = \zeta$, we determine a Taylor series around ζ. This expansion is given by:

$$Z = f(\zeta) + (X_1 - \zeta_1)f_1 + (X_2 - \zeta_2)f_2 + ...$$

$$\tfrac{1}{2}(X_1 - \zeta_1)^2 f_{11} + (X_1 - \zeta_1)(X_2 - \zeta_2)f_{12} + \tfrac{1}{2}(X_2 - \zeta_2)^2 f_{22} + ... , \qquad [6.3a]$$

where f_i denotes the derivative of f according to ζ_i, and f_{ij} the second-order derivative of f according to ζ_i and ζ_j. Both first- and second-order terms should exist. Equation [6.3a] can also be written as:

$$Z = f(\zeta) + (\delta_1 + e_1)f_1 + (\delta_2 + e_2)f_2 + ...$$

$$+ \tfrac{1}{2}(\delta_1 + e_1)^2 f_{11} + (\delta_1 + e_1)(\delta_2 + e_2)f_{12} + \tfrac{1}{2}(\delta_2 + e_2)^2 f_{22} + ... \qquad [6.3b]$$

The second expression is, in particular, appealing as it contains the systematic errors and the random errors. In various cases, we may be able to make various assumptions. The most common assumptions are:

a) No bias occurs, that is, $\delta_i = 0$, then $z = f(\mu)$.

b) The X_i are uncorrelated, hence $E(X_i X_j) = 0$.

c) The probability of large accidental errors is small, hence second- and higher-order terms can be ignored in the Taylor series.

This then brings us to the two famous laws of error propagation:

− The law of the propagation of random errors, valid if all three assumptions hold, equals:

$$var(Z) = f_{11}\, \sigma_1^2 + f_{22}\, \sigma_2^2. \qquad [6.4a]$$

− The law of the propagation of bias equals:

$$\mu_Z \approx f(\mu) + \tfrac{1}{2}(f_{11}\, \sigma_1^2 + f_{22}\, \sigma_2^2). \qquad [6.4b]$$

This law shows that systematic errors (bias) may occur as a consequence of random errors: with $\delta Z \approx \tfrac{1}{2}(f_{11}\, \sigma_1^2 + f_{22}\, \sigma_2^2)$.

A simple example is the area of an ellipse. It is well known that the area of an ellipse is equal to Area $= \tfrac{1}{4}\pi\, \zeta_1\, \zeta_2$. Here, ζ_1 is the length of one axis, and ζ_2 is the length of the other axis. Notice that the area of a circle is a special case, with $\zeta_1 = \zeta_2$. We may then question what the error propagation would be likely to be, given the uncertainty in the two coordinates, ζ_1 and ζ_2. In order to make quantitative statements, we consider the function by $z = f(\zeta_1, \zeta_2) = \frac{1}{4}\pi \, \zeta_1 \, \zeta_2$. We first calculate the two derivatives that are required in equation [6.3a], and find that $f_1(\zeta_1, \zeta_2) = \frac{1}{4}\pi \, \zeta_2$ and that $f_2(\zeta_1, \zeta_2) = \frac{1}{4}\pi \, \zeta_1$. Both derivatives are therefore linear functions in one of the coordinates. Proceeding on this basis, we find that two of the second derivatives are equal to 0, namely the derivatives $f_{11} = 0$ and $f_{22} = 0$, whereas $f_{12}(\zeta_1, \zeta_2) = \frac{1}{4}\pi$, is a constant. This is also the value that emerges for $f_{21}(\zeta_1, \zeta_2)$.

Now suppose that we want to find the area for a few combinations of axis values. First we calculate the area for the axis combination $z = (1,2)$, and suppose that we observe the values $x_1 = 1.1$ and $x_2 = 2.1$. For these values, $f(\zeta_1, \zeta_2) = \frac{1}{2}\pi$ we then find for the area the value of:

$$z = \pi/2 - 0.1 \cdot \frac{1}{4}\pi \cdot 2 - 0.1 \cdot \frac{1}{4}\pi \cdot 1 + 0.1 \cdot 0.1 \cdot \frac{1}{4}\pi$$
$$= \pi/2 - 0.075\,\pi + 0.0025\,\pi \qquad\qquad [6.4c]$$

A similar calculation can be done for other values of ζ_1 and ζ_2, possibly leading to different values of the area and errors in its calculation.

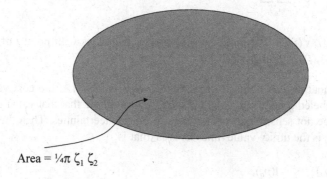

Area $= \frac{1}{4}\pi \, \zeta_1 \, \zeta_2$

Figure 6.3. *The area of an ellipse, as a function of the length two axes ζ_1 and ζ_2*

6.2.5. *The variance-covariance equation*

In a recent paper [VAN 05], the consequences of positional uncertainties for area estimates have been analyzed. The equations that are presented have a general applicability. The variance and covariance in polygon areas are calculated as a function of the variances in the vertex coordinate values. Let $A(p_q)$ be the area of a polygon p_q; the polygon p_q consists of points $(x_q(i),y_q(i))$, $i = 1, \ldots, n_q$, and the polygon p_r of points $(x_r(i),y_r(i))$, $i = 1, \ldots, n_r$. Point i in p_q is labeled with a label $s_q(i)$, point j in p_r with label $s_r(j)$. Points are sorted in counterclockwise direction so that point $i-1$ comes just before, and point $i+1$ just after, point i. On the assumption that these variances are all equal to σ^2, and hence do not depend upon point, polygon, or direction, Van Oort *et al.* [VAN pre] show that var($A(p_q)$) equals:

$$\text{var}(A(p_q)) = \frac{\sigma^2}{4} \cdot \sum_{i=1}^{n_q} \left(\cdot \left(y_q(i+1) - y_q(i-1) \right)^2 + \left(x_q(i+1) - x_q(i-1) \right)^2 \right) + \frac{n_q}{2} \cdot \sigma^4, \qquad [6.5]$$

and that the covariance between $A(p_q)$ and $A(p_r)$ equals:

$$\begin{aligned} \text{cov}(A(p_q), A(p_r)) = \\ \frac{\sigma^2}{8} \cdot \sum_{i=1}^{n_q} \sum_{j=1}^{n_r} \delta(s_q(i), s_r(j)) \cdot \left(y_q(i+1) - y_q(i-1) \right) \cdot \left(y_r(j+1) - y_r(j-1) \right) \\ + \frac{\sigma^2}{8} \cdot \sum_{i=1}^{n_q} \sum_{j=1}^{n_r} \delta(s_q(i), s_r(j)) \cdot \left(x_q(i+1) - x_q(i-1) \right) \cdot \left(x_r(j+1) - x_r(j-1) \right) \\ - \frac{\sigma^4}{8} \cdot \sum_{i=1}^{n_q} \sum_{j=1}^{n_r} \delta(s_q(i), s_r(j)) \left[\delta(s_q(i+1), s_r(j-1)) + \delta(s_q(i-1), s_r(j+1)) \right] \end{aligned} \qquad [6.6]$$

where $\delta(s_q(i), s_r(j)) = 1$ if point i of polygon p_q coincides with point j of polygon p_r and $\delta(s_q(i), s_r(j)) = 0$ otherwise.

This equation can be used in relation to cost values. Let the cost value for the polygon labeled q be given as γ_q; we suppose throughout that that value is given and fixed, hence not sensitive to randomness or other uncertainties. Then the cost of the polygon p_q is the utility value times its area, that is:

$$C(p_q) = \gamma_q \cdot A(p_q), \qquad [6.7]$$

and we find that for the total set of polygons:

$$C(\Omega) = \sum_{q=1}^{N} C(p_q) \cdot \qquad [6.8]$$

Because the cost values are non-random, the variance of the cost value of polygon p_q equals:

$$\text{var}(C(p_q)) = \gamma_q^2 \cdot \text{var}(A(p_q)). \qquad [6.9]$$

By applying the standard expression that the variance of the sum of two random variables equals the sum of the two variances plus the covariance between the two, we find that:

$$\text{var}(C(\Omega)) = \sum_{q=1}^{N} \gamma_q^2 \cdot \text{var}(A(p_q)) + 2 \cdot \sum_{q=1}^{N} \sum_{r>q}^{N} \gamma_q \cdot \gamma_r \cdot \text{cov}(A(p_q), A(p_r)) \,. \ [6.10]$$

It has been shown elsewhere that only the adjacent polygons that have different cost values need to be considered; for the other adjacent polygons, there is no effect of uncertainty on the variance. A simple example of this equation is given in Van Oort et al. [VAN 05], where for two very simple polygons it is explained in detail how errors emerge and how they affect the final calculations. At this stage, it seems to be appropriate to turn toward a real world example, where cost calculations are at hand. Therefore, as an illustration, we now address a case study in the Netherlands.

6.3. Application

6.3.1. Background

Sanitation of contaminated soil is an expensive task. The contaminator can often not be traced to obtain payment, which puts the financial burden on society. Local governments, when quantifying the costs for sanitation, found that estimated costs are sometimes exceeded by 100% [OKX 00]. One of the possible causes is the quality of the input to the cost model. One way to improve upon this is to make cost models as precise and realistic as possible. A well-developed cost model may suffer from low quality input, such as polygons identifying sub-units that can be handled by contractors. The aim of this study was to apply a well-developed cost model to a contaminated area in the industrial part of the Netherlands.

The study area covers a former gasworks site in the harbor area of the city of Rotterdam [VAN 97; BRO 99]. In 1994, integrated soil data was made available, with an observation density to 10 observations per ha, corresponding to a grid of 32m spacing over the whole site. At 99 locations, all contaminants were measured, but at other locations at least two were measured (Figure 6.4). In this study we concentrate on the upper layer (0–0.5 m), which contained the largest set of the full vector of observations. The five contaminants indicative of the spatial extent of contamination are cyanide, mineral oil, lead, PAH-total, and zinc.

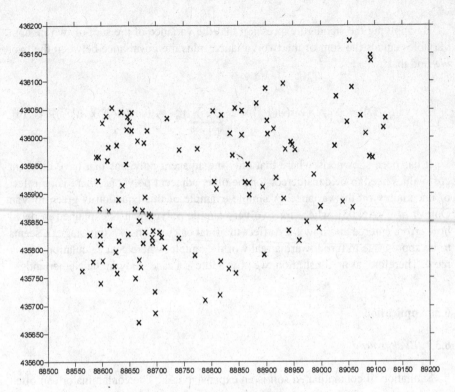

Figure 6.4. *Location of sample points in the area*

Four environmental thresholds classify areas according to the degree of contamination [WBB 97]:

–S-value ['*Safe*' or '*Target level*']: maximum concentration of a contaminant to maintain multi-functionality;

– T-value [=$1/2(S+I)$]: intermediate level between low and serious soil pollution;

– I-value ('*Intervention level*'): if this value is exceeded, the functional properties of the soil in relation to humans, fauna, and flora are seriously degraded or are threatened with serious degradation; and

– BAGA ('*Chemical waste*'): at this level the soil is chemical waste and should be care fully treated.

Table 6.1 shows thresholds for the five contaminants in this study. The S-, T- and I-intervention values are given for a soil profile (10% organic matter and 25% clay). These values are adjusted at individual locations for other organic matter and clay percentages, leading to place-specific threshold values [STE 97].

Contaminant	S-value	T-value	I-value	TG-value	G-value	BAGA
Cn	5	27·5	50	50	50	50
Mineral oil	50	2,525	5,000	500	500	50,000
Pb	85	307·5	530	204	527	5,000
PAH	1	20·5	40	40	40	50
Zn	140	430	720	266	714	20,000

Table 6.1. *Environmental thresholds for 5 contaminants according to two standards. S-, T- and I- values apply to Wet Bodembescherming (Wbb) thresholds, TG-, G- and BAGA values to Interprovinciaal overleg (IPO) thresholds. All thresholds are expressed in (mg kg^{-1})*

Pollution classes for the five contaminants are based upon IPO-thresholds (Figure 6.4) [IPO 97]. These thresholds may differ from Wbb-thresholds (Table 6.1). The IPO-thresholds determine the destination of the excavated soil. IPO-thresholds aim at a controlled, environmentally-justifiable application of the soil, that is, soil with low pollution can still be used at industrial estates, but not in areas that are inhabited. This is in contrast to the standard Wbb-thresholds, which only judge the quality of the soil and lead to sanitation requirements. There are three IPO-thresholds: the S-, the G-, and the TG-threshold with a similar interpretation as the S-, I- and T-thresholds, respectively, but often with different values. An additional threshold exists for soil polluted with inorganic contaminants, the so-called R-threshold, which determines whether soil with inorganic contamination above the G-threshold can still be profitably cleaned by extraction. This only applies when the maximum organic contamination is below the TG-threshold. Further, the BAGA-threshold also applies.

A cost model estimates remediation costs of contaminated sites. Ten classes with different processing costs are distinguished (Table 6.2). To check the necessity for excavation, the degree of pollution is determined separately for all contaminants for all soil cubes, based upon measured or interpolated concentrations (Figure 6.5). The highest degree of pollution of any contaminant determines whether the cube should be excavated or not, depending upon different remediation scenarios.

1a	Non-insulated applicable	6.82
1b	Non-insulated applicable, provided no leaching	9.09
EC	Extractively cleanable	31.77
D	Dump	35.40
DB	Dump BAGA	40.85
TG	Thermally cleanable	31.77
2	ER or dump	34.04
3	TR or dump	34.04
4	ER or TR or dump	34.04

Table 6.2. *Costs (in €) for one 1 m^3 of soil according to different applications*

6.3.2. Results

The area of 282,240 m^2 is divided into 5,760 squares of 7 by 7 m^2. Evaluation of individual squares shows that the treatments 1a, TG and DB occur, with occurrences equal to 2,265, 3,407 and 88 squares, respectively. Applying the cost model gives €3.11 M for treating the total area. Next, to investigate the effects of locational uncertainty, we suppose that the uncertainty in the two directions are the same, and equal to values of 10 cm, 50 cm, and 1 m, respectively. Results are presented in Table 6.3. This gives, as the variance of the total cost, values ranging from 7×10^4 to 7×10^6 euro^2, and of the covariance values ranging from -3×10^4 to -3×10^6 euro^2. Negative values for the covariance are understandable, as an increase in one (say, expensive) part of the area coincides with a decrease in the other (say, cheaper) part of the area. For the three different scenarios, this leads to variance values that range from approximately 4×10^4 to 4×10^6 euro^2. Assuming a Gaussian distribution, a 95% confidence interval as a contribution of the digitized squares to the total costs has a half-length equal to of 200×1.96 € (for $\sigma = 0.1$ m), and to 2000×1.96 € (for $\sigma = 1$ m). The effect is clearly relatively small as compared to the total uncertainty in remediation costs. However, the main benefit is that it can be quantified, using the procedures explained in this study. Other sources of uncertainty may further contribute to the discrepancies between estimated and actual costs [OKX 00].

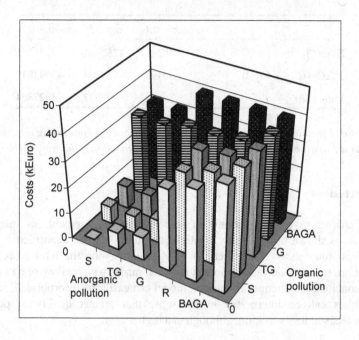

Figure 6.5. *The cost model according to organic and inorganic pollution levels. Vertical bars show the costs for combinations of pollution classes*

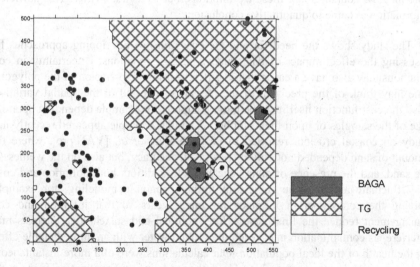

Figure 6.6. *Recommended treatments and sample points in the harbor area*

	$\sigma = 0.1$	$\sigma = 0.5$	$\sigma = 0.1$
$\Sigma Var(C(p_q))$	70296	1762981	7121651
$2\Sigma\Sigma Cov(C(p_q), C(p_r))$	-30492	-765004	-3093847
$Var(C(\Omega))$	39804	997977	4027804

Table 6.3. *Estimated variance of the total costs $U(\Omega)$, and the constituting intermediate results $\Sigma Var(C(pq))$ and $2\Sigma\Sigma Cov(C(pq), C(pr))$. For explanation of the symbols: see the text*

6.4. Conclusions

This chapter gives an overview of some recent developments in spatial data quality. As is shown by this study, spatial data quality can be quantified in various ways. This study shows the effects of polygon quality on the total costs for soil remediation. Some basic assumptions have been made, which allow one to come to precise conclusions. Changes in the cost model can easily be incorporated; results of the analyses can be interpreted in the sense that greater quality of positional accuracy leads to less variation, although at a higher cost.

We have assumed throughout the study that utility values are fixed, that there is no serial correlation in the measurement uncertainty, that uncertainties in the x- and y-direction are the same, and that these uncertainties are independent of location. One may, of course, doubt these assumptions, but as a first step it gives sufficient information to come to quantitative conclusions.

The study shows the need for and the possibility of developing approaches for assessing the effect/impact of data quality on cost functions. Uncertainty in cost functions may also have a cause that is different from the precision of the polygons. One may think of the precision of the measurements and of the spatial variation. Also the cost function itself may be more complex, for example depending upon the size of the samples or upon their number. A similar example appeared recently in a study on coastal erosion, reported by Van de Vlag *et al.* [VAN pre], where the amount of sand depended not only on positional accuracy, but also on the wetness of the sand, and the presence of dune vegetation. In addition, both pollution and costs may fluctuate and may be subject to further economic modeling, for example, leaving the contamination in place for many years so that inflation and cost management reduce the final costs. Here, political and stakeholder interests may interfere, as contamination may reach the ground water with an unpredictable effect on the health of the local population. Our calculations were of a more instantaneous character, allowing the broadening of both the modeling and application scope in the years to come.

6.5. References

[BEL 02] BELHADJALI A., "Using moments for representing polygons and assessing their shape quality in GIS", *Journal of Geographical System*, 2002, vol. 4, p. 209–32.

[BRO 99] BROOS M.J., AARTS L., VAN TOOREN C.F. and STEIN A., "Quantification of the effects of spatially varying environmental contaminants into a cost model for soil remediation", *Journal of Environmental Management*, 1999, vol. 56, p. 133–45.

[CHR 88] CHRISMAN N.R. and YANDELL B.S., "Effects of point error on area calculations: a statistical model", *Surveying and Mapping*, 1988, vol. 48, no. 4, p. 241–46.

[DIL Sub] DILO A. DE BY R. and STEIN A., "Definition of vague spatial objects", *International Journal of Geographic Information Science*, submitted.

[FRA 04] FRANK A.U. and GRUM E. (eds), *Proceedings of the International Symposium on Spatial Data Quality (ISSDQ '04)*, 2004, Vienna, Austria, Department for Geoinformation and Cartography, Vienna University of Technology.

[GOO 98] GOODCHILD M.F. and JEANSOULIN R. (eds), *Data Quality in Geographic Information: From Error to Uncertainty*, 1998, Paris, Editions Hermès Science Publications.

[FRA 98] FRANK A., "Metamodels for data quality description", in *Data Quality in Geographic Information: From Error to Uncertainty*, GOODCHILD M.F. and JEANSOULIN R. (eds), 1998, Paris, Editions Hermès Science Publications, p. 15–29.

[HEU 98] HEUVELINK G.B.M., *Error Propagation in Environmental Modelling with GIS*, 1998, London, Taylor & Francis.

[IPO 97] IPO, Work with secondary materials. Interprovincial policy for application in environmental protection of secondary ingredients (in Dutch), 1997, The Hague, IPO.

[OKX 00] OKX J.P. and STEIN A., "Use of decision trees to value investigation strategies for soil pollution problems", *Environmetrics*, 2000, vol. 11, p. 315–25.

[SHI 03] SHI W., GOODCHILD M.F. and FISHER P.F., *Proceedings of the 2nd International Symposium on Spatial Data Quality 2003*, 19–20 March 2003, Hong Kong, The Hong Kong Polytechnic University.

[STE 97] STEIN A., "Sampling and efficient data use for characterizing polluted areas", in *Statistics of the Environment 3 – Pollution Assessment and Control*, Barnett V. and Turkman K.F. (eds), 1997, Chichester, John Wiley & Sons.

[VAN PRE] VAN DE VLAG D., VASSEUR B., STEIN A. and JEANSOULIN R., "An application of problem and product ontologies for coastal movements", *International Journal of Geographic Information Science*, in press.

[VAN 05] VAN OORT P., STEIN A., BREGT A.K., DE BRUIN S. and KUIPERS J., "A variance and covariance equation for area estimates with a Geographical Information System", *Forest Science*, 2005, vol. 51, p. 347–56.

[VAN 97] VAN TOOREN C.F. and MOSSELMAN M., "A framework for optimization of soil sampling strategy and soil remediation scenario decisions using moving window kriging", in *GeonEnv I*, Soares A., Gomez-Hernandez J. and Froideveaux R. (eds), 1997, Dordrecht, Kluwer, p. 259–70.

[WBB 97] WBB, *Guidelines for Soil Protection, with Accompanying Circulars*, 1997, Staatsuitgeverij, The Hague (in Dutch).

Chapter 7

Reasoning Methods for Handling Uncertain Information in Land Cover Mapping

7.1. Introduction

The object of this chapter is to follow on from Chapter 3 by illustrating how different types of uncertainty may be handled using various formalisms for handling uncertain information. Examples of the uncertainties at the end points of the hierarchy presented in Figure 3.1 are given, using a variety of land cover problems. The hierarchy distinguishes the extent to which the processes under investigation are mappable as spatially distinct objects. The end points in the hierarchy of uncertainty are associated with specific uncertainty formalisms:

- error with probability theory;
- vagueness with fuzzy-set theory; and
- ambiguity with evidential expert approaches.

In this chapter we summarize some of the wider conceptual issues in geographic information uncertainty and present some uncertainty formalisms. We illustrate how these formalisms might be applied, using examples on land cover. Land cover is chosen because it is a phenomenon of interest to many people and a knowledge domain where everyone believes that they are an "expert". For well-defined land cover objects, uncertainties are associated with error. For poorly-defined ones, uncertainties are associated with vagueness and ambiguity. Ambiguity has sub-components, discord and non-specificity, which have been previously described, but for which solutions have not been forthcoming. We present expert evidential

Chapter written by Alexis COMBER, Richard WADSWORTH and Peter FISHER.

approaches that can be applied in different ways to these problems using uncertainty formalisms which are not exclusive to individual sub-components of ambiguity. Rather, expert evidence can be combined using Dempster-Shafer, fuzzy-set theory or endorsement theory, depending on the precise nature of the problem and its context.

7.2. Uncertainty

The uncertainties in geographic information originate from different sources:

– uncertainty due to the inherent nature of geography: different interpretations can be equally valid;

– cartographic uncertainty resulting in positional and attribute errors;

– conceptual uncertainty as a result of differences in "what it is that is being mapped".

Geographic objects are created by humans to impose an order on the physical world. Objects are identified, delineated, and placed into categories according to a set a pre-defined criteria. Whilst many (non-geographic) objects have boundaries that correspond to physical discontinuities in the world (for example, the extent of my cat), this is not the case for many geographic objects that may be less well defined. To explore these uncertainty issues, Barry Smith developed the concept of *fiat* and *bona fide* boundaries, corresponding to *fiat* and *bona fide* geographic objects [SMI 95; SMI 01b; SMI 01a]. Briefly, *fiat* boundaries are boundaries that exist only by virtue of the different sorts of demarcations: they owe their existence to acts of human decision. *Bona fide* boundaries are those boundaries that are independent of human decision. Geographic objects may overlap, and placing an object unequivocally into a particular class may be contentious. Object boundaries will differ from culture to culture, often in ways that result in conflict between groups. Therefore, the boundaries contribute as much to geographic categorical definitions as do the elements that they contain in their interiors [SMI 98].

Typically, a confusion matrix may be used to generate statistics about land cover accuracy, errors of omission, and errors of commission: [CON 91] provides an exhaustive description. The confusion matrix describes cartographic uncertainty due to positional and attribute errors. However, its use is based on the paradoxical assumption that the reference data is itself without error, or at least is superior in quality to the mapped data, although it is difficult to prove that the validation data is more accurate than the one being assessed [CHE 99].

Uncertainty can also arise because of confusion over precisely *what it is that is being mapped*. For example, we may use the term "forest" and be confident that you (the reader) have a clear mental image of what a forest is. You might therefore believe that the notion of what constitutes a "forest" is commonly accepted and is uncontroversial; in which case you will be surprised to discover just how many different ways there are to define a forest. An illustration of this conceptual uncertainty is given in Figure 3.2.

7.3. Well-defined objects: error, probability, and Bayes

Well-defined objects are those where the object classes (for example, land cover class of "fen") are easily separable from other classes and where the individual instances (parcels) are clearly distinct from other instances of the same class. With a well-defined object, the main concerns are positional and attribute errors, and the dominant uncertainty reporting paradigm, the confusion matrix, performs at its best.

A typical problem would be to model the uncertainty caused by errors. For example, the accuracy assessment of the classification of remotely sensed land cover using a referent dataset. In this case Data_1 and Data_2 are two observed land covers. A confusion matrix describes the correspondence of the object class membership, with Data_1 and Data_2 providing identical row and column headings. The pair-wise correspondences provide error measures (user and producer; type I and type II; omission and commission; false positives and false negatives) as described by [CON 91]. These can be used to generate probabilities that the objects in Data_1 are Boolean members of the set of objects in Data_2.

Note that it is quite easy to "misinterpret" what the accuracy derived from the confusion matrix means in practice. Consider the case where there are just two land cover classes (say grass and trees) and that when we test our classifier, 80% of the time that it assesses a point it gets the correct answer. The map suggests that a quarter of the area is grass and the remaining three-quarters are trees; what is the true proportion likely to be? Let us consider what happens if the true areas are 10 pixels of grass and 90 pixels of trees. Of the 10 grass pixels, 8 (10*0.8) will be correctly identified and 2 will be incorrect; of the 90 pixels cover in trees, 72 (90*0.8) will be correctly identified and 18 will not. Because we have assumed that our classification system is complete, we will determine that there are 26 pixels of grass (8 correct and 18 misclassified trees). So despite our classifier being rather good (80% is very respectable), we have over-predicted the amount of grass by a factor of three and the probability that any pixel identified as grass really is grass is 0.31, not 0.80! Conversely, although we will underestimate the area of trees, our confidence that a pixel identified as being trees is trees is very high at 0.97 (72 correctly identified and the 2 misidentified grass pixels, probability = 72/74).

The obvious solution to this is to adopt a "Bayesian" approach where the objective is to determine the extent to which we wish to *revise our belief* in a "hypothesis" in the light of new evidence.

Bayesian probability is:
$$P(h \mid e) = \frac{P(e \mid h).P(h)}{\Sigma P(e_i \mid h_i).P(h_i)}$$
[7.1]

where:

$P(h|e)$ posterior probability that hypothesis "h" is true given the evidence "e";

$P(h)$ prior probability that hypothesis "h" is true;

$P(e|h)$ probability of observing evidence "e" when hypothesis "h" is true and the subscript "i" indicates all competing hypothesis (assumed to be exhaustive).

Consider the data in Table 7.1. Data_1 and Data_2 are two independent and contemporary land cover surveys from remotely sensed data. From Data_1 we have an initial set of probabilities (*prior probabilities*) to describe the likelihood that a pixel is a member of each of the 5 classes:

P({wood, arable, urban, grass, water}) = {0.21, 0.28, 0.2, 0.22, 0.09}

		Data_2					
		Arable	Grasslands	Urban	Water	Woodland	Total
	Woodland	1	0	5	0	15	21
	Arable	19	2	4	1	2	28
Data_1	Urban	2	1	13	1	3	20
	Grasslands	2	11	5	2	2	22
	Water	0	0	1	7	1	9
	Total	24	14	28	11	23	100

Table 7.1. *Example land cover correspondence data*

If Data_2 is considered as the reference data or ground truth, then the prior probabilities (X) for each class can be modified by this new data (Y). For example, for woodlands, the prior probability of a pixel of woodlands in Data_1 being found to be woodlands in Data_2 can be modified, where our alternative hypothesis is that it is not woodland:

$P(h_{wood}) = 21/100$

$P(e|h_{wood}) = 15/21$

$P(h_{\neg wood}) = 79/100$

$P(e|h_{\neg wood}) = 8/79$

$$P\left(\text{Woodland}_{\text{Data\_1}} \mid Woodland_{Data\_2}\right) = \frac{(15/21) \times (21/100)}{((15/21) \times (21/100) + (8/79) \times (79/100))} = 15/23$$

That is, the pair-wise correspondence divided by the total from the referent data (column). We might have been able to intuitively identify this accuracy term. However, this was for a straightforward question: what is the probability of a pixel of *Woodland* in Data_1 being woodland in Data_2? If a more revision-specific question is asked, the answer is less straightforward. Consider the problem of change where Data_1 and Data_2 represent land cover data collected at two times, t1 and t2, and a pixel of *Woodland* in Data_1 is found to be *Urban* in Data_2. In this case, taking a Bayesian perspective, the answer depends not just on how good our classifier is, but on the relative frequency of the two classes.

The question that the Bayesian approach is answering is *"to what extent has my belief in X changed because of this data?"*, as expressed by the unconditional probability that A is true, given evidence "Y". This approach is best applied to uncertainty problems where precise probabilities are available for all circumstances. Some care is needed in the application of Bayes where a full probability model may not be available, and even when it is, it may hide some dangerous assumptions. For instance, the prior probabilities in the example above are weighted by area/frequency of occurrence. These prior probabilities may not reflect the process under investigation (for example, land cover change through urbanization) and tend to weight the outcomes to the class of object with greatest already-observed frequency. However, probabilistic approaches are used in many applications that lack a complete probability model. This is because the rules of probability calculus are uncontroversial; they ensure that the conclusions are constant with the probability assessments and it is easy to understand what the model does. This may be misleading, and in practice it is difficult to make precise assessments of probabilities.

Probabilistic approaches are most suited to problems where there are probabilities for all events and least suited to problems where there is partial or complete ignorance, or limited or conflicting information. They cannot deal with imprecise, qualitative, or natural language judgments, such as "if A then possibly B".

7.4. Poorly-defined objects: spatial extent, vagueness, and fuzzy-set theory

Vagueness occurs for poorly-defined objects (class or instance); where it is difficult to identify the spatial extent or to precisely define an object, it is difficult to allocate unequivocally individual objects into a class. That is, it is not possible to define the spatial extent of the object to be analyzed and no combination of attributes or properties allows the object to be allocated to a class. This results in vagueness which can be modeled using fuzzy-set theory.

For example, consider a land cover classification of remotely sensed imagery. The classic approach is to treat the spectral information as independent dimensions and characterize each "class" in terms of its central location in this spectral space. A pixel is assessed in terms of how far it is from the central location of each class; often this distance is adjusted to take into account how compact or diffuse a particular class is. [RIC 93] describes the process in detail. The most likely class for a pixel to belong to is the one that it is closest to. For example, consider the hypothetical data in Table 7.2 of distances to each of 5 land cover class centers of 4 pixels. Each pixel is classified by determining the class center to which the pixel is closest (that is, the shortest distance away). For some pixels, the allocation of a land cover class label is unproblematic: pixel 3 is very close to the woodland class prototype. However, in other cases the differences between alternative pixel labels is small; there is more uncertainty associated with the allocation of the label Urban to pixel 1, given its closeness to Woodland.

		Relative distance to class centers				
	Most likely attribution	**Arable**	**Grass**	**Urban**	**Water**	**Woodland**
Pixel 1	Urban	0.872	0.872	0.530	0.983	0.598
Pixel 2	Urban	0.978	0.843	0.180	1.000	1.000
Pixel 3	Woodland	1.000	0.902	0.918	0.951	0.230
Pixel 4	Woodland	0.966	0.805	0.690	0.977	0.598

Table 7.2. *Relative distances to class prototypes for example pixels*

Fuzzy approaches calculate the membership function of each object (for example, a pixel) to all possible classes. The data in Table 7.2 can be reworked by subtracting the distance from unity to get the fuzzy membership functions for each pixel for each class in Table 7.3. It is then possible to visualize the fuzzy map as a series of layers, each one relating to a particular class. This process is illustrated in Figure 7.1, where the fuzzy membership for each pixel (in this case 1km^2) is derived by aggregating the proportion of different classes from a finer spatial resolution dataset.

	Pixel label	Fuzzy membership functions				
	Pixel label	Arable	Grassland	Urban	Water	Woodland
Pixel 1	Urban	0.128	0.128	0.470	0.017	0.402
Pixel 2	Urban	0.022	0.157	0.820	0.000	0.000
Pixel 3	Woodland	0.000	0.098	0.082	0.049	0.770
Pixel 4	Woodland	0.034	0.195	0.310	0.023	0.402

Table 7.3. *Fuzzy membership functions for example pixels*

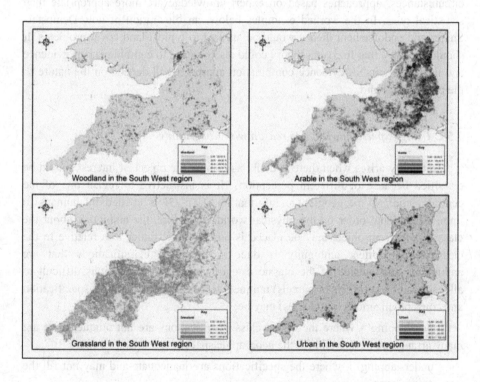

Figure 7.1. *Fuzzy maps of land cover for the South West region in the UK, from the ADAS land database*

7.5. Poorly defined specification: ambiguity, discord, non-specificity and expert knowledge

7.5.1. *Ambiguity*

Ambiguity is composed of discord and non-specificity. Discord arises when one object is clearly defined, but could be placed into two or more different classes under differing schemes or interpretations of the evidence [FIS 99]. Non-specificity occurs when the assignment of an object to a class is open to interpretation [FIS 99]. In the case of ambiguity, neither the probabilistic (Bayesian) nor fuzzy approaches are efficient; this is because the correspondence matrix does not adequately capture the discordant and non-specific relations between two datasets. In these circumstances, approaches based on expert knowledge are more appropriate than statistical ones. In the worked examples below, an illustration of how Dempster-Shafer and endorsement theory would be applied to typical land cover problems. It should be noted that fuzzy-set theory could also be used to combine expert evidence and that the choice of evidence combination approach will depend on the nature of the problem at hand.

7.5.2. *Using expert knowledge to reason with uncertainty*

Ambiguity arises when there is doubt as to how the object of interest should be classified because of different perceptions. It is instructive to revisit one of the central paradigms of geographic information (GI): GI is created by humans to impose a rational order on the physical world. Therefore, the ambiguity about the class in which an object may be placed is not absolute, rather it is relative to the classification. Often, ambiguity is due to individual classifications that are ambiguous and outside of the classic categorization constraints: it is difficult to allocate an object (pixels or parcels) uniquely to one class, as the class specification and conceptualizations (ontologies) may be:

– overlapping – where the object class specifications are not unique, there are many arrangements that will fill the geographic space;

–'under-lapping' – where the specifications are inadequate and may not fill the geographic space.

Formalizing expert knowledge offers a method to overcome such differences in object specification.

Expert or knowledge-based approaches are increasingly being used to relate discordant spatial information [COM 04a; COM 04b; SKE 03; FRI 04]. An expert approach is able to take the knowledge that is embedded within different maps and their construction. It offers a more holistic view of the data and the landscape processes that may be differently represented. The use and need for expert systems is increasing as the number and availability of spatial data and geographic information systems (GIS) users increases. GI straddles many disciplines – botany, ecology, remote sensing, planning, environmental management, etc. Each discipline has its own disciplinary paradigms and issues that shape its analytical approaches. The interaction of these are difficult to formalize, but might be quite easily articulated in natural language by a domain expert.

7.5.3. Formalisms for managing ambiguity

7.5.3.1. Discord, experts and Dempster-Shafer theory of evidence

Discord occurs where an individual object is *bona fide*, but could be allocated to more than one class from different classification schemes. Both the UK land cover mappings of 1990 and 2000 include classes of bog: Upland Bog and Lowland Bog target classes in the 1990 LCMGB and the Bog broad habitat of LCM2000. One might quite reasonably expect there to be some relation between the two definitions of bog. This is not the case. The impact of the differences in conceptualization of bog between the two surveys is illustrated in Table 7.4, which contains the definitions of bog from LCMGB and LCM2000. It is evident that they are conceived in very different ways: LCMGB by standing water, and LCM2000 by peat depth. An uninformed user comparing the two different datasets might conclude that there is a strong relation between the two conceptualizations of bog; further, that in this area (the Ordnance Survey 100x100 km square SK) there has been a considerable increase in bog over the past ten years, despite the inertia of bog vegetation. Therefore, there is a situation where the datasets are nominally very similar (both classify Landsat satellite data to map land cover and both include a class Bog), but they are difficult to compare or to integrate because they embody different views of the landscape.

Land Cover Mappings	'Bog' Key Features	Area of Bog
1990 LCMGB from http://www.ceh.ac.uk/ data/lcm/classM.htm	*Permanent waterlogging,* resulting in depositions of acidic peat. Mostly herbaceous communities of wetlands with *permanent or temporary standing water,* Lowland bogs: carry most of the species of upland bogs, but in an obviously lowland context, with *Myrica gale* and *Eriophorum* spp. being highly characteristic. Upland bogs: have many of the species of grass and dwarf shrub heaths. Characterized by *waterlogging, perhaps with surface water,* especially in winter. Species such as bog myrtle *(Myrica gale)* and cotton grass (*Eriophorum* spp.), in addition to the species of grass and dwarf shrub moorlands.	< 1ha
LCM2000 from http://www.ceh.ac.uk/data/ lcm/lcmleaflet2000/leaflet3.pdf	Bogs include ericaceous, herbaceous, and mossy vegetation in areas with *peat >0.5 m deep*; ericaceous bogs are distinguished at subclass level. Inclusion of ericaceous bogs contrasts with LCMGB 1990, where bogs were herbaceous or mossy in seasonal standing water.	7,550 ha

Table 7.4. *Discord between the LCMGB and LCM2000 definitions and mapped areas of bog in the OS tile SK, taken verbatim from the indicated sources and with the salient features underlined by the authors*

An expert, who is familiar with both datasets and the differing conceptualizations, could be used to identify relations between, for example, the LCM2000 class of bog and LCMGB classes. The authors have developed an approach, using Dempster-Shafer, to model expert opinion on how the classes in two datasets relate to each other [COM 04b], [COM 04c], [COM 05] in order to identify inconsistencies (that is, relative errors or actual changes) between datasets. It is illustrated in the worked example below.

An expert has described the relations between elements of two classifications, basing the description on their expert opinion. The relations are expressed in a three-valued logic that describes pair-wise relations that are Anticipated, Uncertain, Improbable {A, U, I}. The pair-wise relations can be visualized as in Table 7.5; note that this represents a many-to-many relationship; a class in a system (map) is related to all classes in the second system (as anticipated, uncertain or improbable).

		Map_1			
		A	**B**	**C**	**D**
	X	1	-1	-1	0
Map_2	**Y**	0	-1	1	0
	Z	-1	1	0	1

Table 7.5. *Hypothetical expert relations between elements of two classifications*

Consider a hypothetical LCM2000 parcel of class X that has been intersected with LCMGB and generates the following distribution of LCMGB pixels:

Class A, 53 pixels

Class B, 7 pixels

Class C, 11 pixels

Class D, 24 pixels.

Using the relations in Table 7.5, it is possible to generate values for the set {A, U, I} by summing the number of pixels corresponding to each relation: {53, 18, 24}. The expert has expressed uncertain, as well as positive and negative, relations in the table. These lend themselves to analysis using the Dempster-Shafer theory of evidence.

Dempster-Shafer is the only theory of evidence that allows for the explicit representation of uncertainty. It requires a pair of functions to assess a proposition: the belief function (Bel) and the plausibility function (Pls). The uncertainty (Unc) is the difference between the plausibility and belief. The belief is therefore a lower estimate of the support for a proposition and the plausibility an upper bound, or confidence band. Disbelief (Dis) is 1- Pls. They are related in the following way:

$$Bel + Dis + Unc = 1.$$ [7.2]

The greater the anticipated score, the higher the belief that the parcel has not changed since 1990. Evidence from other sources (for example, "A" and "B") can be combined using the Dempster-Shafer theorem:

$$Bel_{AB} = \frac{Bel_A Bel_B + Bel_A Unc_B + Bel_B Unc_A}{\beta}$$ [7.3]

where β is a normalizing factor that ensures Bel + Dis + Unc =1, and

$$\beta = 1 - Bel_A Dis_B - Bel_B Dis_A \qquad\qquad [7.4]$$

Dempster-Shafer assigns a numerical measure to the evidence (mass assignment, m) over sets of hypotheses, as well as over individual hypotheses. Dempster–Shafer is not a method that considers the evidence hypothesis by hypothesis, rather it assesses the evidence for related questions. Whereas Bayesian approaches assess probabilities directly, Dempster-Shafer answers the question "what is the belief in A, as expressed by the probability that the proposition A is provable given the evidence?" [PEA 88].

Dempster–Shafer can model various types of partial ignorance, limited or conflicting evidence, Dempster–Shafer is most suited to situations where beliefs can be expressed numerically, but there is some degree of ignorance, that is, there is an incomplete model. Dempster-Shafer is a suitable method for combining or analyzing expert information and for reasoning about uncertainty when there is ambiguity caused by discordant classifications or interpretations of the data. Other work exploring the limits of Dempster-Shafer theory for combining expert evidence in a GIS context includes [TAN 02; COM 04b; COM 04d and FER 97].

7.5.3.2. *Non-specificity, experts, and qualitative reasoning formalisms*

Non-specificity occurs when the allocation of an object to a class is open to interpretation. In these circumstances expert opinion is needed to define a set of rules or beliefs that allow allocation decisions to be made. Endorsement theory is a non-numeric approach developed by [COH 85], and has been used in some automated mapping applications where different types of evidence have been combined (for example, [SRI 90; SRI 93; SKE 97; COM 02; COM 04a; COM 04d]. The allocation of geographic objects to classes can be a subjective process. Qualitative reasoning formalisms require that four aspects of the problem be defined [SUL 85] and [COH 85]:

1. The different strengths of expert belief must be identified and named.

2. The interaction of beliefs when combined must be specified to produce overall endorsements.

3. A system for ranking endorsements must be specified.

4. Qualitative thresholds of belief must be defined to decide when the evidence accrued is "believable enough".

For example, consider a parcel of land cover, previously mapped as class X and which is suspected to have changed. The strengths of belief in different types of evidence can be defined as follows:

– *definite* if a single piece of evidence indicates that the hypothesis is true (note that this kind of *evidence* is rare in the identification of land cover change direction);

– *positive* if the evidence supports the hypothesis, but may be contradicted;

– *average* if the evidence contributes some support for the hypothesis;

– *none* if the evidence contributes no support for the hypothesis.

Similarly, the interaction of beliefs can be specified to generate hypothesis endorsements:

– *certain* if the evidence provides definite belief and no definite disbelief;

– *believed* if the combined evidence provides positive belief and no positive disbelief;

– *plausible* when average belief is greater than average disbelief;

– *conflicting* when the weights of belief and disbelief are equal.

From expert knowledge we have a set of possible change directions, given its previously mapped land cover, and beliefs in different types of evidence for each of the hypothetical change directions, as in Table 7.6. From the expert, there are descriptions of how important the different types of evidence are, relative to the specific pair-wise changes (note that none of the evidence for the hypothesized change direction is *definite*). For example, it might be that for changes from X to A, soil quality information is more important than for changes from X to D. Real examples are given by [SKE 97], detecting felled forestry where the most important information is the change in spectral signal combined with spatial rules, and by [COM 04a], considering semi-natural land cover change, where many types of evidence are combined.

Class X Evidence	Change direction				Change Area
	A	**B**	**C**	**D**	
Change in spectral signal	Strong (*average*)	Weak (*positive*)	Weak (*positive*)	Very strong (*average*)	Weak
Soil type	Poor (*positive*)	Rich (*positive*)	Rich (*average*)	Good (*none*)	Rich
Slope	Steep (*average*)	Gentle (*none*)	Steep (*average*)	Very steep (*positive*)	Steep

Table 7.6. *Evidence for different change directions from an original land cover mapped as X, the strength of belief, and the characteristics of that area of change. Matches in the characteristics are highlighted*

In the candidate area, the change in spectral signal is weak, the soil type is rich, and the land steep. For hypothesis "A", this data provides average support for one criteria (the slope). In the same way, the other alternatives can be evaluated.

It is then possible to evaluate each hypothesis according to the scheme for combining beliefs:

– hypothesis *A* has only one set of average belief from the evidence, and is *plausible*;

– hypothesis *B* has two sets of positive belief, and is *believed*;

– hypothesis C has two sets of average belief and one of positive belief, and is *believed*;

– hypothesis *D* is conferred no belief by the evidence.

In this case, hypotheses *B* and *C* are both believed, but whilst the 'mass of evidence supports *C*, the weight of evidence is greater for *B* with two sets of positive belief. Although the example includes only a limited number of evidence types, change directions, and belief endorsements, it does illustrate the basic application of endorsement theory.

The endorsement model takes a much more heuristic approach to reasoning about uncertainty than the other approaches. It allows the definition of beliefs and their interaction to be specified according to the problem being considered, and the question that it addresses is "what are the sources of uncertainty in the reasoning process, and where were they introduced?" The meaning of the answer is then interpreted through the method by which endorsements combine and how the endorsements are ranked. It has a number of advantages [COM 04d]: First, it is able to represent common knowledge (such as expert mapping rules) in a natural from. Secondly, its symbolic approach can represent and reason with knowledge for real-world problems. Thirdly, reasoning in this way allows inferences to be drawn from partial knowledge [SRI 93]. Fourthly, the results of endorsement approaches contain explicit information about why one believes and disbelieves. Consequently, it is possible to reflect on these, and decide how to act – a very useful property for the subjectivity of mapping.

Endorsement-based approaches are most suited to situations where subjective degrees of belief do not generally behave as probabilities or are not numerically expressed. The knowledge elicitation phase of the construction of expert systems is one such application area: domain experts are often uncomfortable committing themselves to numerical values. However, they may be inappropriate for domains in which numerical degrees of belief have a clear semantics and are adequate expressions of all information about uncertainty.

7.6. Conclusion

All information is uncertain to some extent whether we acknowledge it or not. GI is no exception and the very nature of GI means that it ought to be considered. In this chapter we have described the main ways that uncertainties arise in geographic information. Which form of uncertainty is the most important depends on the nature of the process under investigation and the objective of the study; that which is measured and recorded in a dataset may not be universally accepted by all: different algorithms, interpreters, techniques, and object conceptualizations will result in different mapped information from the same raw data. Thus, the implications of not understanding in the fullest sense what it is that is being mapped are profound in terms of the results of any analyses.

The definitional issue is further developed by exploring the problem of defining geographic objects: any abstraction of reality results in information loss (and therefore uncertainty). The nature and direction of that information loss may not be apparent from the data and is compounded by classification issues around *well-defined* objects (for example, land ownership, cadastral units) and *poorly-defined* objects (for example, zones of transition between different semi-natural land covers). The philosophical constructs of *fiat* and *bona fide* boundaries, corresponding to *fiat* and *bona fide* geographic objects, illustrate that the difficulties of placing a geographic object unequivocally into a particular class may also be contentious because defining many geographical objects necessarily involves an arbitrary drawing of boundaries in a continuum, boundaries that differ from culture to culture.

The aim of this chapter has been twofold. First, to illustrate the diversity of uncertainty that is present in much data and analysis, even within the discipline of land cover. Secondly, to show how the different types of uncertainty can be treated in different ways, using different approaches. To do this we have focused on different types of land cover problems and how they require different types of solution, depending on the nature of their uncertainty. In this way we have gone beyond that which is usually described in the GI literature. Typically, descriptions of uncertainty approaches are given with a focus on the theoretical models, without illustration of their application to actual GI problems. These descriptions have tended to describe what can be measured, rather than the true uncertainties associated with the problem at hand and have ignored the very real problem of discord and non-specificity. Although not widely used, there are tools to cope with ambiguity and we hope this chapter has gone some way to introduce them.

7.7. References

[CHE 99] CHERRILL A.J. and MCCLEAN C., "Between-observer variation in the application of a standard method of habitat mapping by environmental consultants in the UK", *Journal of Applied Ecology*, 1999, vol 36, number 6, p 989–1008.

[COH 85] COHEN P.R., *Heuristic Reasoning About Uncertainty: An Artificial Intelligence Approach*, 1985, Boston USA, Pitman Advanced Publishing.

[COM 02] COMBER A.J., Automated land cover change, 2002, PhD Thesis, University of Aberdeen, Scotland.

[COM 04a] COMBER A.J., LAW A.N.R. and LISHMAN J.R., "Application of knowledge for automated land cover change monitoring", *International Journal of Remote Sensing*, 2004, vol 25, number 16, p 3177–92.

[COM 04b] COMBER A.J. FISHER P.F. and WADSWORTH R.A., "Integrating land cover data with different ontologies: identifying change from inconsistency", *International Journal of Geographic Information Science*, 2004, vol 18, number 7, p 691–708.

[COM 04c] COMBER A.J., FISHER P.F. and WADSWORTH R.A., "Assessment of a semantic statistical approach to setecting land cover change using inconsistent data sets", *Photogrammetric Engineering and Remote Sensing*, 2004, vol 70, number 8, p 931–38.

[COM 04d] COMBER A.J., LAW A.N.R. and LISHMAN J.R., "A comparison of Bayes', Dempster-Shafer and endorsement theories for managing knowledge uncertainty in the context of land cover monitoring", *Computers, Environment and Urban Systems*, 2004, vol 28, number 4, p 311–27.

[COM 05] COMBER A.J., FISHER P.F. and WADSWORTH R.A., "Identifying land cover change using a semantic statistical approach", in *Geodynamics*, Atkinson P., Foody G., Darby S. and Wu F. (eds), 2005, Boca Raton, CRC Press, p 73–86.

[CON 91] CONGALTON R.G., "A review of assessing the accuracy of classifications of remotely sensed data", *Remote Sensing of Environment*, 1991, vol 37, number 1, p 35–46.

[FER 97] FERRIER G. and WADGE G., "An integrated GIS and knowledge-based system as an aid for the geological analysis of sedimentary basins", *International Journal of Geographical Information Science*, 1997, vol 11, number 3, p 281–97.

[FIS 99] FISHER P.F., "Models of uncertainty in spatial data", in Longley P.A., Goodchild M.F., Maguire D.J. and Rhind D.W. (eds) *Geographic Information Systems*, 2nd edn, 1999, vol 1, New York, John Wiley & Sons, p 191–205.

[FRI 04] FRITZ S. and SEE L., "Improving quality and minimising uncertainty of land cover maps using fuzzy logic", *Proceedings of the 12th Annual GIS Research UK Conference*, Lovatt A. (ed), 28–30 April 2004, UEA, Norwich, UK, p 329–31.

[PEA 88] PEARL J., *Probabilistic Reasoning in Intelligent Systems: Networks of Plausible Inference*, 1988, San Mateo, Morgan Kaufmann.

[RIC 93] RICHARDS J.A. and JIA X., *Remote Sensing Digital Image Analysis*, 1993, Berlin, Springer-Verlag.

[SKE 97] SKELSEY C., A system for monitoring land cover, 1997, PhD Thesis, University of Aberdeen, Scotland.

[SKE 03] SKELSEY C., LAW A.N.R., WINTER M. and LISHMAN J.R., "A system for monitoring land cover", *International Journal of Remote Sensing*, 2003, vol 24, number 23, p 4853–69.

[SMI 95] SMITH B., "On drawing lines on a map", *Spatial Information Theory: Lecture Notes in Computer Science*, 1995, vol 988, p 475–84.

[SMI 98] SMITH B. and MARK D.M., "Ontology and geographic kinds", *Proceedings of the 8th International Symposium on Spatial Data Handling (SDH'98)*, Poiker T.K. and Chrisman N. (eds), 12–15 July 1998, Vancouver, Canada, p 308–20.

[SMI 01a] SMITH B. and MARK D.M., "Geographical categories: an ontological investigation", *International Journal of Geographical Information Science*, 2001, vol 15, number 7, p 591–612.

[SMI 01b] SMITH B., "Fiat objects", *Topoi*, 2001, vol 20, number 2, p 131–48.

[SRI 90] SRINIVASAN A. and RICHARDS J.A., "Knowledge-based techniques for Multi-Source Classification", *International Journal of Remote Sensing*, 1990, vol 11, number 3, p 505–25.

[SRI 93] SRINIVASAN A. and RICHARDS J.A., "Analysis of GIS spatial data using knowledge-based methods", *International Journal of Geographic Information Systems*, 1993, vol 7, number 6, p 479–500.

[SUL 85] SULLIVAN M. and COHEN P.R., "An Endorsement based Plan Recognition Program", *Proceedings of the 9th International Joint Conference on Artificial Intelligence*, Aravind J.K. (ed), August 1985, Los Angeles, USA, Morgan Kaufmann, p 475–79.

[TAN 02] TANGESTANI M.H. and MOORE F., "The use of Dempster-Shafer model and GIS in integration of geoscientific data for porphyry copper potential mapping, north of Shahr-e-Babak, Iran", *International Journal of Applied Earth Observation and Geoinformation*, 2002, vol 4, number 1, p 65–74.

Chapter 8

Vector Data Quality: A Data Provider's Perspective

8.1. Introduction

With increasingly diverse uses and users for geographical information, together with growing accessibility to global positioning technologies and applications that integrate datasets, the provision of meaningful data quality information, along with geographical data products, is arguably ever more important. Whether one is a specialist user, or casual consumer user, of data products, there is a need to understand the suitability and limitations of data in the context in which it will be used, that is, context of use in space and time. For example, the developing demands for real-time information services elevate, for some, the significance of temporal quality in geographical data.

This chapter aims to give perspectives on quality in vector geographical data and ways in which the data is managed from the position of a data provider, that is, Ordnance Survey, the national mapping agency of Great Britain. It covers the importance of understanding user needs, structure of a national vector dataset, application of data quality elements in the capture through to product supply process, and communication of data quality information to users of the data.

Chapter written by Jenny HARDING.

8.2. Providing vector geographical data

8.2.1. *Data quality and usability*

From a data provider's perspective, data quality is the degree to which data in a dataset conforms to the capture specification and to product specifications for the dataset. These specifications describe the provider's representation or abstraction of the real-world, sometimes referred to as the "universe of discourse" [ISO 02], "abstract universe", or "terrain nominal". From a customer or end-user perspective, concern about data quality is more focused on the fitness for purpose of the data for their intended use or application. It is important, too, that the data provider understands as far as possible the uses to which customers may wish to put the data in order to evaluate fitness of specifications and suitability of information provided to customers on data quality to help customers decide on fitness for their purpose. These provider and user perspectives may be referred to respectively as "internal quality" and "external quality". The overall quality of a geographical information product is dependent on the standard of both internal and external quality.

Given the enormous diversity of potential uses for geographical information, it is not feasible for a data provider to fully understand the needs of all potential users of its data products, but it remains important for the interests of both provider and user that the quality of data products is known and described in meaningful terms. From the perspective of a national mapping agency as a data provider, the specifications for data capture and eventual data products largely reflect its organizational responsibilities and objectives.

8.2.2. *Aims of a national mapping agency*

Ordnance Survey is a government agency responsible for the definitive surveying and topographic mapping of Great Britain and for maintaining consistent national coverage of its geographical information datasets. From its beginnings in 1791 as a provider of military mapping, Ordnance Survey has developed with advances in technology, organization, and customer needs into the geographical information business it is today.

Its primary aim is to satisfy the national interest and customer need for accurate and readily available geographical information and maps in Great Britain. Commitment to improving geographical information content, delivery, and fitness for purpose, and meeting the current and future needs of customers are central to Ordnance Survey's objectives [ORD 04a]. In order to achieve this aim, attention to data quality throughout the data capture to data supply process is of critical importance.

8.2.3. *Vector geographical data*

In vector geographical data, the real world is modeled and represented geometrically and topologically by points, lines, and polygons (areas). These data are enriched through association of attributes and text. The Ordnance Survey dataset for large-scale vector data is held in an object-based database from which products for customer supply are derived. A brief overview of this dataset is given here in the context of the derived OS MasterMap™ product, as background information to later sections in this chapter which focus on aspects of vector data quality. For the purposes of this chapter, the focus is on planimetric vector data (data in the horizontal x,y dimensions), and height data is not specifically referred to.

OS MasterMap is a framework for referencing geographical information in Great Britain [ORD 04b], and is comprised of vector data in themes of feature groupings and positioned relative to the British National Grid. It also includes an ortho-rectified raster imagery layer. The data form a seamless representation of the whole country, with the basic data unit being the "feature". Features composed of points, lines, or polygons represent real-world entities in accordance with an abstraction of the real-world encapsulated in a data capture specification. Included are topographic features, such as buildings, roads, watercourses, and non-topographic features, such as administrative boundaries and postal addresses. A complete list of real-world topographic entities and their representation in OS MasterMap can be found in the "OS MasterMap real-world object catalogue" [ORD 04c]. Consistent with the concept of a digital national framework [ORD 04d], every feature has a unique identifier called a TOID®, which remains the same throughout the feature's life. A version number is also applied as an attribute of each feature and incremented each time there is any change to the feature in the database.

8.3. Data quality needs of the end user

8.3.1. *Users' understanding of their needs*

Users' needs for data content and the quality of that content are rooted in the purpose for which the data are to be used. Geographical information is often an information source contributing to problem-solving or decision-making; hence, reliability of outcomes is, in part, based on the fitness for purpose of the data source itself as well as on its interoperability with other data sources. Users therefore need to consider whether a data source will provide them with the type and quality of information needed in the context of intended use. For example, are relevant real-world features located to a sufficient level of accuracy; are required spatial, temporal, and thematic attributes present and consistent across the area of interest;

are connectivity relationships in the data suitable for analysis required; are the data up-to-date enough for the purpose?

To help potential users determine whether geographical data are fit for the intended purpose, the UK Association for Geographic Information, for example, publishes guidelines [AGI 96] that focus attention on what questions the user is trying to answer, coverage and currency required, and so on. The guidelines also recommend that users think about what assurance they require that the data are to their required specification in content and quality. For example, as pointed out by Morrison [MOR 95], computer renderings of digital data often give the appearance of accurate information, regardless of the data's actual accuracy.

With growth in wireless and internet services, however, many new users of geographical information are not selecting information sources directly themselves, but rather indirectly as part of a service. While these users may not have, or even want, access to specification and quality information about integrated data sources, their basic question is the same. Does the service provide the right information, when and where it is needed, to serve the purpose? The fitness for purpose of source data must be decided by service developers and providers, with service success partly dependent on usability, reliability, timeliness, content, and quality of output presented to the user.

8.3.2. *Data providers' understanding of user needs*

Success as an information provider depends largely on providing information that is fit for the purpose for target users. National mapping organizations provide geographical information for increasingly diverse uses in professional and everyday life, and understanding these needs is a key priority. To investigate evolving user needs, Ordnance Survey facilitates complementary lines of enquiry and user feedback. These include targeted market research questionnaires, consultations with user groups, understanding gained through customer account management, and responding to customer feedback. Looking to emerging new uses, as well as enhancing information for existing applications, analytical research focuses on the user, task context, and technological opportunities. Here the aim is to identify what geographical information is needed in terms of content and quality, when, where, and how, to assist the user in their decision making. This research informs future development of capture, database, and product specifications, as well as ways to communicate geographical information effectively to the end user.

8.4. Recognizing quality elements in vector data

An overview of data quality in the capture to supply process is given in section 8.5. This section explains standard elements of spatial data quality in the context of how they apply in Ordnance Survey vector data. Reference to data quality elements may vary between data providers. Results from a survey[1] of 20 European and one American national mapping agencies suggest some differences in the way that quality elements are applied, and, notably, just 62% reported using positional accuracy as a data quality element [JAK 01].

A quality element is a measurable, quantitative component of information about data quality [ISO 02]. Data quality for Ordnance Survey large-scale vector data is described and controlled with respect to five quality elements: currency (temporal validity), positional accuracy, attribute accuracy (thematic accuracy), logical consistency, and completeness [ORD 04b]. In addition, the non-quantitative component, lineage, provides further information about data quality. Starting with lineage, these components of quality information are described here, together with their acceptable levels of conformance in Ordnance Survey large-scale vector data.

8.4.1. *Lineage*

Lineage describes the life history of a dataset. That is, the sources, methods, and processes used in the creation and maintenance of the dataset, from real-world abstraction and capture through to representation in digital data. Given Ordnance Survey's long history of large-scale mapping, the lineage of current vector data is complex. Digital data capture from drawn large-scale master survey documents began in 1971, and has since been maintained through a number of field- and office-based capture and editing processes which have developed with advances in technology. Furthermore, data capture and vector feature coding specifications have undergone a series of changes; the most recent major change being transformation of the entire large-scale dataset to a link and node structure with features in themed layers, building greater intelligence into the data and forming the basis of OS MasterMap [ORD 04b].

For these reasons, full lineage information for points, lines, and polygons in the dataset is not recorded. Further information is, however, available at feature level showing the date the feature geometry or attribution was last updated, update status (indicating the type of change applied to a feature), change history (recording superseded update date and update status values), and the acquisition method for x,y and z coordinates of each data point. The acquisition method refers to capture

---

1 Survey by Eurogeographics Sub-Working Group on Data Quality, 1999 [JAK 01].

methods, described in section 8.5.2, from which expected positional accuracy can be inferred. Quality control measures for lineage information within the dataset are implemented through validation software which checks the integrity of data field values and syntax against the current data specification.

8.4.2. *Currency*

Currency is a parameter of how up-to-date the data are, and can be viewed as a form of semantic accuracy; that is, a parameter of how well a real-world object is actually described in the data [SAL 95]. Other forms of semantic accuracy include attribute accuracy, completeness, and consistency, described later in this chapter.

From a data producer's perspective, currency, or "temporal validity" [ORD 04b], is a parameter of data capture and update policy. Ordnance Survey's update policy for large-scale vector data distinguishes between two categories of change. The first concerns features of significant business or national interest, including areas of housing, commercial or industrial developments, and communications networks. These are captured within six months of construction under a "continuous revision" policy. The second category includes features of generally lower benefit for customers, such as vegetation and small building-extensions. These are captured within a cyclic revision process every five years, or ten years in mountain and moorland areas [ORD 02].

Achievement of these currency targets can only be as good as the organization's knowledge of real-world change. Measurement of conformance is dependent upon quantifying and comparing known real-world change with captured change in the data. Detecting and quantifying real-world change relies on a combination of local knowledge, third-party information sources, and, potentially in the future, automated change detection from imagery. While striving toward the ultimate goal of real-time change detection and capture, it presently remains difficult for a national mapping organization to achieve 100% currency. Levels of conformity therefore aim to meet or exceed published "acceptable quality levels". For example, in 2004, Ordnance Survey's acceptable quality level for major change in urban areas was that, on average, no more than 0.6 units of change of over six months old per 0.25 square km remains unsurveyed[2]. This quality level translates into an overall agency performance monitor, expressed as a target percentage of real-world change to be represented in the national database within six months of construction [ORD 04e]. What is considered the date of construction will of course differ depending on

---

2 Ordnance Survey quantifies real-world change in terms of "units of change", allowing, for example, the degree of change incurred by a section of new road to be quantified in units equivalent to an area of new housing development. For definitions see, for example, the OS MasterMap User Guide [ORD 04b].

perspective of interested parties (e.g. construction date of a house could be understood as the date construction work started or completed).

Within the captured vector dataset itself, each feature is attributed with the date it was last updated, though this does not necessarily inform the user about currency. For example, a building feature last updated in 1950 may still be up to date with respect to the real-world and current update policy. From a user's perspective, it is important to note that the date of update of a vector feature relates to the supplier's capture policy and usually will not coincide with the date at which change actually occurred in the real world. This difference could be a significant factor in the assessment of quality or fitness for purpose of the data in some specific user-applications.

8.4.3. Positional accuracy

Positional accuracy may be defined as the degree to which the digital representation of a real-world entity agrees with its true position on the earth's surface. Absolute accuracy of Ordnance Survey vector data coordinates (that is, planimetric x,y coordinates) is measured against the actual position of the real-world entities in relation to the British National Grid reference system, as defined by the OSGB36 triangulation. Based on accuracy-tested results of sampled data over the past 30 years, absolute accuracy of large-scale vector data is represented by root mean square error (RMSE)[3] values in relation to the scale of survey, as shown in Table 8.1.

Survey type	Root mean square error	95% confidence level	99% confidence level
1:1,250 (urban)	± 0.42 m	± 0.73 m	± 0.90 m
1:2,500 resurvey (rural/urban)	± 1.10 m	± 1.90 m	± 2.40 m
1:2,500 overhaul (rural/urban)	± 2.70 m	± 4.67 m	± 5.79 m
1:10,000 (mountain and moorland)	± 4.09 m	± 7.08 m	± 8.78 m

Table 8.1. *Absolute accuracy of Ordnance Survey vector data*

---

3 The RMSE value represents here the discrepancy between the captured position of real-world entities and their true position in relation to the British National Grid reference system.

Two further components of positional accuracy are assessed for Ordnance Survey vector data, namely geometric fidelity and relative accuracy. The latter is the positional consistency of a data point in relation to other local data points. To measure relative accuracy, scaled distances between sampled data points on well-defined features (for example, building corners) are compared with distance measured between these points on the ground. Results are expressed in terms of "expected standard error vector" applicable to the map area. Geometric fidelity is the "trueness" of features in data to the shapes and alignments of real-world entities they represent [ORD 04b]. Alignments that are straight in the real-world, for example, must be represented as straight alignments in the data. Acceptance of conformity to quality requirements is a matter of visual judgment when the data are viewed at a scale no larger than source-survey scale, and is initially made at the point of data capture.

Reported errors in positional accuracy are addressed during the normal update cycle, or earlier if warranted by their impact on customers' businesses. The current policy for maintaining positional accuracy in vector data is to ensure that new detail is surveyed to a positional accuracy with RMSE not exceeding 0.4 m in 1:1,250 scale areas, 1.0 m in 1:2,500 scale areas and 3.4 m in 1:10,000 scale areas.

Initial capture method and subsequent post-capture processing determine the positional accuracy of vector data that is ultimately supplied to the user. Such lineage information can be invaluable to positional error estimation in some end-user applications [DRU 95]. In fact, due to the cumulative effects of past vector data lineage on absolute positional accuracy, together with growing expectations for improved positional accuracy through increasingly widespread use of global positioning technologies, a number of mapping organizations have embarked on major positional accuracy improvement programs. Often these programs are associated with moving national or regional mapping data to a new map projection system, as for example in Austria, Bavaria, Ireland, and Northern Ireland [RÖN 04a, RÖN 04b], or localized adjustment to control points.

A past transformation of this kind is a key factor underlying the program for positional accuracy improvement of Ordnance Survey large-scale vector data for rural areas. Topographic mapping for rural areas was originally based on a transformation in the 1950s from the County Series mapping, where each county had its own projection and coordinate system, to the British National Grid system. While the relative accuracy of the data was good (± 1.2 m RMSE), limitations in the conversion process resulted in an absolute accuracy of around ± 2.7 m RMSE [ORD 04f]. The positional accuracy improvement program aims to benefit both Ordnance Survey and its customers through future-proofing the database for the addition of new building developments and providing better positional interoperability between rural vector data and customers' own data positioned by global positioning systems (GPS). Where customers already integrate their own positional information with topographic vector data, improvement to positional accuracy does entail a process of migration, which is an issue that requires careful consultation between data provider and customer [RÖN 04c].

8.4.4. *Attribute accuracy*

Attribute accuracy is the accuracy to which attributes in the vector-data record information about real-world entities. Included here are attributes of feature representation, such as feature classification, text information for feature names or descriptions, and change history attributes at feature level. Measures of conformance are made in terms of the percentage of correct attributes of a given type within a sample of data.

While attribute accuracy can be complex and subject to uncertainty of, for example, measurement and interpretation [GOO 95], levels of uncertainty in capturing attributes in topographic data capture and in subsequent testing can to some extent be minimized through clear definition in capture specifications, operator training, and logic-based quality control measures built into data capture and editing systems.

8.4.5. *Logical consistency*

Logical consistency is a measure of the degree to which the data logic and syntax complies with the data structures defined in the dataset specifications. Quality control for logical consistency in Ordnance Survey large-scale vector data includes checks for topological consistency, validity of record structure, and validity of values. For example, connectivity is checked to make sure polygons are "closed" properly (aspects of vector connectivity are discussed in more detail below). The validity of individual attribute values and combinations of attributes form a check on content and context logic. Checks are completed automatically by software, ensuring that values of feature classifications and attributes, geometry and topology, database schema, and file formats are valid in accordance with the data specification.

8.4.5.1. *Connectivity in vector data*

Structuring rules determine the connectivity of features in vector data. Some key rules are summarized here. A line feature, representing a fence for example, is composed of two XY coordinate tuples for the ends of each straight segment. Where the end of a line feature intersects or terminates within a small, specified distance of another line feature in the same structuring layer, the end must "snap" to the other line feature; that is, there must be no overshoots or undershoots. Junction coordinates are recorded for the intersection (see Figure 8.1). Where an end of line feature does not connect to another feature, the "free ends" must be verified. Some real-world features, such as posts, are represented by points, as are certain symbols and area labels classifying bounding polygons (denoting, for example, roofed areas, water, or vegetated areas). In order for polygon classifications not to "leak" outside the polygon, polygons must be "closed" by the first and last coordinate tuple being identical (Figure 8.1). Coordinates for lines bounding adjacent polygons (for example, for a fence dividing two field polygons) are stored once to avoid slivers or overlaps and gaps arising from slight differences in boundary coordinates, as an artifact of digital capture. Further connectivity establishes network topology between certain features, such as the center lines of roads. The above and other structuring rules form an essential part of quality control checks (see section 8.5.1.2).

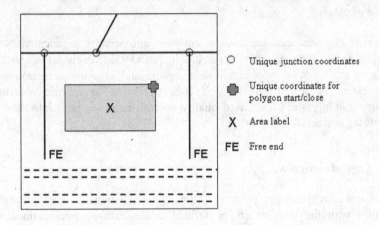

Figure 8.1. *Vector line and polygon connectivity*

8.4.6. *Completeness*

From a data supplier's perspective, completeness is a measure of the degree to which data content corresponds to the real world in accordance with the data capture specification, dataset coverage, and at the level of currency required by the update policy. Non-conformance may be measured in terms of omission, where entities in the real world are covered by the capture specification, but do not appear in the data, and commission, where features remain in the data after they have disappeared in the real world, or where features exist in the data, but do not conform to the capture specification.

8.5. Quality in the capture, storage, and supply of vector data

8.5.1. *Overview of the data capture to supply process*

Ultimately, the quality of data supplied to the customer is, in most part, only as good as the quality of original data capture. Having said this, certain aspects of data quality can be improved post-capture, as explained in section 8.5.3. Once captured, data are processed through a number of systems before being supplied to the customer in data products, and at each stage, data quality must be checked and maintained through quality control and quality assurance measures. This section provides an overview of that capture to supply process, highlighting key data-quality considerations.

A simplified view of the process is given in Figure 8.2, showing the capture, storage, and delivery stages, with quality control (QC) an integral part of all stages and quality assurance as an independent check on quality at stages of the process. These checking functions are further explained below. Producing good quality vector data begins, however, with the definition of the capture specification, which also provides the key reference for subsequent quality checks.

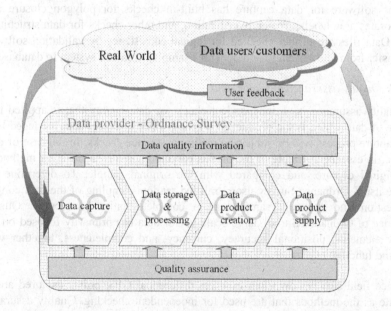

Figure 8.2. *User needs, data capture and supply process*

8.5.1.1. *Capture specifications*

Large-scale vector data are captured in accordance with controlled data-specification documentation aimed at maximizing consistency in the content, structure, and quality of national coverage. This documentation defines an abstraction of the real world, based on customer need and the responsibilities of a national mapping agency. Important components included in the specification are: descriptions and diagrams of data structure; feature classification and definitions for all attribute fields; rules for consistency in, for example, selection of what real-world features are to be represented and how to represent them; algorithms that are to be applied; principles in the collection of data; and data quality standards applying to the data specified. Defined data quality standards provide the basis for post-capture quality tests and statements on resultant data quality.

8.5.1.2. *Quality control of vector data*

The term "quality control" encompasses checks applied to data at all stages of the data capture, editing, storage, and manipulation process, to ensure that data integrity is maintained. Quality control measures are built into system processes so that non-conformance with data specifications is either prevented from occurring or identified and corrected as close to the source as possible. For example, editing-systems software for data capture has built-in checks for polygon closure and incorporates area labels, free-end verification, and other checks for data structuring logic. Data files are further checked for logical consistency by validation software on transfer from one system to another (for example, from edit system to database).

8.5.1.3. *Quality assurance of vector data*

Quality assurance is an independent check for data quality that is applied to a sample of data which is the product of a process. For instance, data produced by a data capture process will be subject to quality assurance checks. In the case of data capture, in essence a sample of the data is captured again by a different method to the original capture, and compared with the original sample. To determine the sample itself, Ordnance Survey uses stratified random sampling of the data content in question and normally sampled from complete national coverage. Quality assurance of Ordnance Survey large-scale vector data has primarily focused on the quality elements, positional accuracy, currency, and completeness, together with form and function of data.

Since field survey now often involves differential GPS, points captured are as accurate as the methods that are used for independent checking. Quality assurance tests for positional accuracy are currently focused on data received from third-party suppliers of photogrametrically-captured data. Selected points in supplied data are checked for absolute and relative accuracy in the field; geometric fidelity is also checked.

Quality assurance for currency and completeness produces the agency performance metric referred to in section 8.4.2. For this, a statistically significant sample of data is taken from the national dataset, where the sample data represent urban, rural, mountain, and moorland areas, but the sample data is weighted to urban areas where feature density is higher. Areas covered by the sample are perambulated in the field, any non-conformance with the specification and update policy is recorded, and information on omissions is fed into a change-intelligence system database, identifying items for future surveys. A national currency result is calculated with respect to the density of real-world features expected in the sampled data.

8.5.2. *Quality in data capture*

A combination of field survey, photogrammetry, and external data sources is used to capture large-scale vector data, with the choice of method taking into account a number of technical, economic, data quality, and safety factors. Of these factors, data quality metrics for positional accuracy are key criteria for determining capture method options for a given location. RMSE of existing data, together with the size of site to be surveyed, determines the need to provide survey control points by GPS, instrumental, or photogrammetric survey in order to achieve required positional accuracy.

Each capture method is considered below with respect to data-quality constraints, but before that, a comment on precision in position measurement: while some measurement and positioning technologies can provide coordinates to centimeter precision or higher, it is important to distinguish between precision of digital measurement and the degree of precision that is meaningful with respect to the geographical representation [VAU 02]. In other words, it may be meaningless to make calculations, for example, based on coordinates in meters to greater than one decimal place in precision, when the data capture specification requires representation of geographical entities to 0.2 m resolution. Planimetric coordinates for all feature geometry in Ordnance Survey's large-scale vector data are supplied to the nearest 0.01 m, as it is necessary to represent absolute accuracy levels achievable by the most accurate GPS in current use. It should be noted however, that this apparent precision in coordinates does not mean that all points in the data are accurate to 0.01 m. An acquisition method attribute applied to each data point indicates by what method that data point was captured and therefore what absolute accuracy RMSE is expected for the point.

8.5.2.1. *Field capture of vector data*

Much of the requirement for new survey control points and, increasingly, the positioning and geometry capture for new real-world entities are provided through use of GPS technologies. This improves overall absolute accuracy of resultant vector data and thereby improves interoperability on a positional level with other GPS-based datasets. Captured coordinates are transformed by software to the OSGB36 coordinate system of the British National Grid.

Field data capture may also involve other instrumental survey techniques, such as electronic distance measurement. A detailed survey, capturing the position and thematic attribution to represent real-world objects to the full capture specification, is completed using graphical survey. This technique positions new detail relative to the positions of existing detail in vector data, using a combination of distance measurement and lines of sight within a set of defined tolerance levels. The once inevitable degradation of field survey positioning, caused by post-capture digitizing of detail, is eliminated by direct capture of digital position coordinates into a mobile field-editing system for vector data.

Classification and attribution of positioned feature geometry is also completed using the field-editing system, which incorporates software to validate input values as far as possible. This built-in validation enables a level of quality control to be provided at the point of data capture, detecting errors before data reach the maintenance database.

8.5.2.2. *Photogrammetric capture of vector data*

Photogrammetric surveys from ortho-rectified, high-resolution aerial imagery account for a significant proportion of vector-data production, especially for rural, mountain, and moorland areas. Field completion provides the additional data required to meet the full capture specification.

In common with many national mapping organizations [JAK 01], Ordnance Survey contracts out part of its photogrammetric production work in order to accelerate data capture programs. Specialist contractors, both within and outside the UK, work to the same capture specifications and quality standards as in-house photogrammetric revision programs, as stipulated in contract documentation. Contractors, of which a growing proportion has accredited processes[4], are responsible for the quality control of their production work to ensure quality acceptance levels are met. Supplied data are then acceptance tested to ensure conformance with specification and required quality levels.

8.5.2.3. *External sources for vector data*

Some new developments, such as large areas of new housing, are initially captured into large-scale vector data from developers' digital site-plan documents.

---

4 Ordnance Survey accredits contractors' processes to ensure that those processes consistently produce data of the required quality.

After a process of registering the data to existing detail control points, cleaning the data to Ordnance Survey specification, and verifying the data through field tests, the data are accepted into the vector dataset.

8.5.3. *Quality in the storage and post-capture processing of vector data*

All captured data are processed through validation software before databanking which checks that all values for data fields are valid in terms of data logic and syntax. With a number of data capture and edit systems requiring access to the maintenance database and operating concurrently in office and field locations, it is also necessary to control access to data *from* the database for update and maintenance. For this reason, "write" access for the update of data is controlled so that non-conformances, in, for example, currency, are not introduced by wrongly sequencing data extractions and update returns to the database.

Since automated quality controls do not yet detect or require corrective action at source for all errors, there will be some items of data stored in the national database that do not conform to the data specification. When detected, these "errors" are categorized and recorded in a database of "defects", a system for the monitoring of non-conformance and the assessment of corrective action required. Non-conformance occurs when the number of defects in the data exceeds the acceptable quality level for the quality element in question. In this way, "acceptable quality level" may alternatively be expressed as "maximum non-conformance", the limiting value for a quality metric. Known errors requiring field resolution are identified to the surveyor, via the field editor, whilst the surveyor is working in the vicinity.

Driven by the need to improve data products for users, some types of errors and some specification enhancements are addressed through quality improvement programs. These programs involve amending data in the entire national database for the specific improvement required. Examples include ensuring that all descriptive text conforms to specification and is correctly associated to vector features.

8.5.4. *Quality in vector data product creation and supply*

Before new products or new product versions are released and are available from the customer supply database, automated quality control checks for logical consistency are applied when loading data from the maintenance database to a test supply database. Any errors are automatically logged and resolved (automatically where possible) before re-loading. Once accepted, logical consistency checks are again applied on transfer of data to a live supply database.

Finally, a random sample is taken of data extracted for customers from the supply database, and looked at as the customer would look at it. This involves putting the data through format translation products and loading data into systems typically used by customers. In this way the "customer inside" can detect and act on any discrepancies, with respect to the product-user guide, which could affect actual end users.

8.6. Communication of vector data quality information to the user

Where use of data is in a decision support context, the user may be faced with multiple uncertainties in the decision-making process, of which uncertainty about fitness for purpose of geographical information may be one with lesser or greater impact on decision outcomes. Furthermore, spatial data may be just one source of information to be integrated with other data sources. Information about data quality needs to be unambiguous and enable the user to minimize, as far as possible, their uncertainty when first deciding on the fitness of the data for their purpose and, secondly, when interpreting data output. In this way, data with meaningful and reliable specification and quality information are of more value to the user.

Two principal information sources about spatial data products can help the user. The first are metadata services summarizing information about datasets' accessibility, contents, formats, and so on. Some data quality information may be included here, such as lineage and currency. Examples include the GIgateway Data Locator, a metadata search engine for UK geospatial datasets. The second source is that provided directly by the data producer or supplier. Information describing data content and quality should be available to check before acquiring a data product, and should also be supplied with the data. Quality statements are for the benefit of data users and should, therefore, be relevant and accessible to users, quantitative wherever possible, easily understood, and attributable in terms of how the quality statistics or values were derived [SMI 93].

Documents on the Ordnance Survey web site that describe OS MasterMap include the real-world object catalogue, revision policy, information on supply file format, and the product-user guide. The product-user guide incorporates statements on acceptable quality levels for completeness, positional accuracy, temporal accuracy, logical consistency, and attribute accuracy applying to the national dataset. Statements on metrics for specific quality parameters by region or local area are not currently given with product information. Once data are acquired by the user, however, further quality information is embedded in data attributes at feature level, for example, positional accuracy information, and can be accessed by interrogating the data.

Survey results suggest there is strong consensus among national mapping agencies recognizing the need for proper data quality specification and for educational material concerning data quality specifications and assurance routines [JAK 01]. The same survey particularly highlighted the need for better quality evaluation processes and communication of data quality evaluation results.

8.7. Conclusions and future directions

Data quality is a fundamental consideration throughout the process of producing vector geographical data, starting with user needs and finishing with communication of data quality information to the end user. This chapter has aimed to show how acceptable quality levels and quality controls are integral to the production process and influence decisions throughout the process. With reference to quality standards for geographical information, the data quality elements, lineage, currency, positional accuracy, attribute accuracy, logical consistency, and completeness, provide key parameters that require critical attention by both data provider and data user.

With increasing demands for real-time data services, greater interoperability of data in applications and services, together with national and international drives towards spatial data infrastructures (for example, Infrastructure for Spatial Information in Europe (INSPIRE)), the importance of both data providers and end-users understanding implications of data quality is becoming ever more important. These trends point to a need for further examination by data providers of what quality levels meet different user needs and how best to evaluate and communicate data quality information in ways that are meaningful to producer and user. This, coupled with improved awareness by service providers and end users of how fit for purpose data are in the context of use, can serve to build confidence in, and wider benefits of, geographical information usage.

Challenges for the future from a data provider's perspective might therefore focus on vector data quality in various ways, including the following: analysis of diverse user-needs for geographical information; identification of what is needed, when, and to what levels of quality in the context of use; cost effective expansion of automated quality-control measures at the point of data capture, helping the achievement of a faster, more efficient, capture to supply process; enhanced quality information at feature level, in a seamless data environment; improved data quality evaluation processes, and communication of meaningful evaluation results, be they mediated by service providers or direct to the professional or consumer user.

8.8. References

[AGI 96] ASSOCIATION FOR GEOGRAPHICAL INFORMATION, *Guidelines for Geographic Information Content and Quality: For Those who Use, Hold or Need to Acquire Geographic Information*, 1996, London, AGI publication number 1/96.

[DRU 95] DRUMMOND J., "Positional accuracy", in GUPTILL S.C. and MORRISON J.L. (eds) *Elements of Spatial Data Quality*, 1995, Oxford, Elsevier, p. 31–58.

[GOO 95] GOODCHILD M.F., "Attribute accuracy", in GUPTILL S.C. and MORRISON J.L. (eds) *Elements of Spatial Data Quality*, 1995, Oxford, Elsevier, p. 59–79.

[ISO 02] INTERNATIONAL ORGANISATION FOR STANDARDIZATION, *ISO 19113:2002 Geographic Information – Quality Principles*, 2002, Geneva, ISO.

[JAK 01] JAKOBSSON A. and VAUGLIN F., "Status of data quality in European national mapping agencies", *Proceedings of the 20th International Cartographic Conference*, 2001, Beijing, China.

[MOR 95] MORRISON J.L., "Spatial data quality", in GUPTILL S.C. and MORRISON J.L. (eds) *Elements of Spatial Data Quality*, 1995, Oxford, Elsevier, p. 1–12.

[ORD 02] ORDNANCE SURVEY, *Information Paper: Revision Policy for Basic Scale Products*, 2002.
http://www.ordnancesurvey.co.uk/oswebsite/products/landline/pdf/BasicScalerevisionpolicy.pdf

[ORD 04a] ORDNANCE SURVEY, Framework Document 2004.
http://www.ordnancesurvey.co.uk/oswebsite/aboutus/reports/frameworkdocument2004.pdf

[ORD 04b] ORDNANCE SURVEY, OS MasterMap User Guide Product Specification v.5.2, 2004.
http://www.ordnancesurvey.co.uk/oswebsite/products/osmastermap/pdf/userguidepart1.pdf

[ORD 04c] ORDNANCE SURVEY, OS MasterMap Real World Object Catalogue, 2004,
http://www.ordnancesurvey.co.uk/oswebsite/products/osmastermap/pdf/realWorldObject Catalogue.pdf

[ORD 04d] ORDNANCE SURVEY, The Digital National Framework – evolving a framework for interoperability across all kinds of information, Ordnance Survey White Paper, 2004.
http://www.ordnancesurvey.co.uk/oswebsite/aboutus/reports/dnf_white_paper.pdf

[ORD 04e] ORDNANCE SURVEY, Annual Report and Accounts 2003–04, London, The Stationery Office, 2004.

[ORD 04f] ORDNANCE SURVEY, Positional Accuracy Improvement – background information, 2004.
http://www.ordnancesurvey.co.uk/oswebsite/pai/backgroundinformation.html

[RÖN 04a] RÖNSDORF C., "Improve your business position", GITA conference, April 2004, Seattle, USA.

[RÖN 04b] RÖNSDORF C., "The Pole Position – high accuracy geodata and geometric interoperability in Great Britain, Europe and the US", *Geomatics World*, 2004, vol. 12, no. 6, p. 22–24.

[RÖN 04c] RÖNSDORF C., "Positional integration of geodata", *Intergeo East conference*, March 2004, Belgrade.

[SAL 95] SALGÉ F., "Semantic accuracy", in GUPTILL S.C. and MORRISON J.L. (eds), *Elements of Spatial Data Quality*, 1995, Oxford, Elsevier, p. 139–51.

[SMI 93] SMITH N.S. and RHIND D.W., "Defining the quality of spatial data: a discussion document", in MASTERS E.G. and POLLARD J.R. (eds), *Land Information Management, Geographic Information Systems*, 1993, vol. 1, *Proceedings of the LIM & GIS Conference*, Sydney, p. 139–51.

[VAU 02] VAUGLIN F., "A practical study on precision and resolution in vector geographical databases", in SHI W., FISHER P. and GOODCHILD M.F. (eds), *Spatial Data Quality*, 2002, London, Taylor & Francis, p. 127–39.

Acknowledgements

I would like to thank Ray Patrucco, Steve Cowell, Phil Mogg, Andy Tidby and Jonathan Holmes for their discussions on data quality handling by Ordnance Survey and for comments on the text, for which I also thank Doreen Boyd and Ed Parsons. © Crown copyright 2005. Reproduced by permission of Ordnance Survey.

Disclaimer

This chapter has been prepared for information purposes only. It is not designed to constitute definitive advice on the topics covered and any reliance placed on the contents of this chapter is at the sole risk of the reader.

Chapter 9

Spatial Integrity Constraints: A Tool for Improving the Internal Quality of Spatial Data

9.1. Introduction

Many geospatial data users tend to think that data quality refers to data planimetric accuracy and structure (compliance with rules governing base primitives, such as surface closure). Other aspects, however, can also significantly influence the overall quality of a geographic dataset. One such case is the spatial integrity of geographic data, which is usually described with spatial relations. Spatial relations can be divided into two categories: metrical relations and topological relations. Metrical relations deal with the measurable characteristics of objects (for example, distance in time or space, directions), whereas topological relations concern the geometric properties that remain invariable after transformation of reference objects (for example, overlaps, disjoint).

This chapter deals specifically with topological relations and explains how they can be used to ensure the spatial integrity of geographic data, thereby improving the data internal quality. Consider, for example, data related to a theme dealing with hydrology. Beyond the requirements of planimetric accuracy and compliance with geometric primitives, a user might want a dataset in which the islands are located within the lakes, the rapids are actually in the rivers, and the roads do not cross the waterbodies (Figure 9.1). This desirable state of data could be described as spatial

---

Chapter written by Sylvain VALLIÈRES, Jean BRODEUR and Daniel PILON.

integrity constraints. To illustrate, the spatial integrity constraints below, presented as axioms, determine the expected state of the data:

– Roads cannot cross waterbodies.

– Islands must lie totally within waterbodies.

– Islands must be disjoint from each other.

Figure 9.1 provides an example of geometric inconsistency between two objects because they have distinct meanings (road and lake). In principle, therefore, a line may cross a polygon (left) without engendering inconsistencies. If the line represents a road and the surface a lake (right), however, spatial inconsistency exists between the two objects.

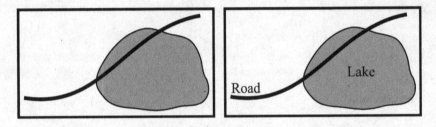

Figure 9.1. *Example of spatial inconsistencies between a road and a lake*

Such inconsistencies can be caused during data acquisition (for example, error in entering the object's code or generalization of an object resulting in inconsistency with another one) or following integration of geospatial data from different sources or of different levels of accuracy. Validation of spatial integrity constraints makes it possible to minimize entry errors and data integration inconsistencies in order to produce a set of integrated data that conform to the stated constraints.

This chapter also presents a way of documenting and describing spatial integrity constraints between different geographic objects based on existing international standards in geomatics. Moreover, it offers additions to deal with certain aspects that are not covered by these standards.

We begin by presenting an overview of work related to the topic of this chapter (section 9.2). Section 9.3 provides a review of the international standards dealing with the spatial integrity of geographic data. Section 9.4 describes the concepts proposed to describe spatial integrity constraints, as well as the additions to existing international standards. Section 9.5 provides the formal representation of spatial integrity constraints and their use in a context for producing geographic data, more specifically, the Centre for Topographic Information (CTI), an agency of Natural Resources Canada (NRCan). The use of spatial integrity constraints is illustrated through an example in section 9.6. The chapter ends with a conclusion in section 9.7.

9.2. Existing work

The literature contains a number of documents dealing with the development, documentation, and use of spatial integrity constraints. More specifically, they pertain to topological relations, controlling the quality of geographic databases, managing the spatial integrity constraints on geographic data, data dictionaries, and so on.

Topological relations enable the description of the state between two spatial objects *a* and *b*, both *a priori* and *a posteriori* (that is, before data entry and in the data *per se*). In order to formalize the spatial relations between two geometric objects, Egenhofer proposed a four-intersection model comparing the boundaries and interiors of two geometric objects[1] [EGE 93; EGE 91; EGE 95], followed by a second model with nine intersections, which includes intersections with the exteriors of geometric objects [EGE 97; EGE 94a; EGE 94b]. These models are represented by intersection matrices. Figure 9.2 and Table 9.1 provide descriptions of the notions of interior, boundary, and exterior for the geometries *point*, *line*, and *surface*, as well as the nine-intersection matrix that applies to these various geometries.

Each matrix intersection takes the value T (true) or F (false), depending on whether the intersection is non-empty ($\neg\emptyset$) or empty ($\emptyset$). A certain number of predicates are defined, based on the values of the different intersections in the matrix: disjoint, overlaps, within, equals, contains, crosses, and touches.

1 The four-intersection model was developed specifically using polygon objects.

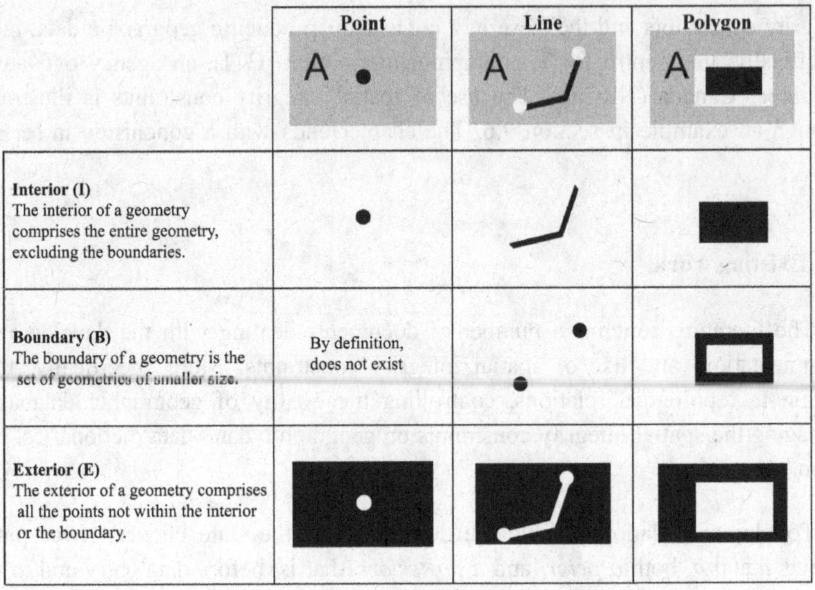

	Point	Line	Polygon
Interior (I) The interior of a geometry comprises the entire geometry, excluding the boundaries.			
Boundary (B) The boundary of a geometry is the set of geometries of smaller size.	By definition, does not exist		
Exterior (E) The exterior of a geometry comprises all the points not within the interior or the boundary.			

Figure 9.2. *Concepts of interior, boundary, and exterior pertaining to geometric primitives*

	Interior *(b)*	**Boundary** *(b)*	**Exterior** *(b)*
Interior *(a)*	$I(a) \cap I(b)$	$I(a) \cap B(b)$	$I(a) \cap E(b)$
Boundary *(a)*	$B(a) \cap I(b)$	$B(a) \cap B(b)$	$B(a) \cap E(b)$
Exterior *(a)*	$E(a) \cap I(b)$	$E(a) \cap B(b)$	$E(a) \cap E(b)$

Table 9.1. *Egenhofer's 9-intersection matrix*

Clementini and Di Felice went further, proposing the *calculus-based method* to express the topological relation between two geometric objects and extend the initial model to nine intersections [CLE 95; CLE 96]. This method uses similar predicates in addition to providing operators to qualify intersection size when it is non-empty ($\neg\varnothing$). They use the value 0 for a point intersection (⊡²), 1 for a line intersection (⊠), and 2 for a surface intersection (⊟) (see Table 9.2).

---

2 The pictograms representing point, line and surface features are documented in [BED 89].

Value	Meaning	Explanation
T	True	The intersection is non-empty. There is an intersection, i.e. $x \neq \varnothing$
F	False	The intersection is empty. There is no intersection, i.e. $x = \varnothing$
*	All values	The intersection can be empty or not.
0	True	The intersection is non-empty. There is an intersection., i.e. $x \neq \varnothing$ The resultant geometries contain at least **one point** (⊡)
1	True	The intersection is non-empty. There is an intersection, i.e. $x \neq \varnothing$ The resultant geometries contain at least **one line** (◩)
2	True	The intersection is non-empty. There is an intersection., i.e. $x \neq \varnothing$ The resultant geometries contain at least **one polygon** (▣)

Table 9.2. *Possible values for Clementini and Di Felice's intersection matrix*

In an attempt to improve the quality of geographic databases, Hadzilacos and Tryfona [HAD 92] proposed a method to express spatial integrity constraints in a database based on Egenhofer's four-intersection model.

Ubeda [UBE 97a; UBE 97b] enriched this approach by defining spatial integrity constraints as 13 subsets of topological relations (for example, equal, disjoint, touches, borders) using Clementini's dimensionally extended nine-intersection model. He then developed management and constraint-validation methods, as well as a tool for correcting errors in existing geographic data.

Cockcroft [COC 97] developed a taxonomy of spatial integrity constraints that apply to geographic data and recently proposed a model for a repository [COC 04] for the management of spatial data integrity constraints.

Parent *et al.* [PAR 97] proposed conceptual data modeling using pictograms to manage spatial integrity constraints.

Normand [NOR 99] proposed a generic approach to define and represent spatial integrity constraints. This method, which is also based on Egenhofer's nine-intersection model, was implemented in a tool that has been used to document spatial integrity constraints for two digital map products belonging to the Quebec provincial government (BDTQ and BDTA).

Borges [BOR 02] offered an extension of the object modeling technique (OMT) model for geographic applications similar to the *Plugin for Visual Language* [BED 99] used in *Perceptory*. The purpose of the research is to illustrate the object classes with their spatial representations in the OMT language [RUM 91] to express integrity constraints between object classes.

Recently, Mostafavi *et al.* [MOS 04] used the Prolog (Programming in Logic) language to manage spatial integrity constraints and the data governed by them. They tested their approach with data from the National Topographic Data Base (NTDB) of Canada. The approach provided the means for (1) checking the consistency between integrity constraints, and (2) automatically checking data consistency with respect to the constraints.

9.3. Topological relations and international geomatics standards

ISO/TC 211 is the International Organization for Standardization (ISO) technical committee responsible for developing international standards in geomatics. Many international standards have already been published; others soon will be. International standards in geomatics covers topics related to spatial and temporal representation, application schemata, data dictionaries, the description and assessment of data quality, metadata, services related to geographic data, and so on. Two standards deal more specifically with defining spatial integrity constraints: Spatial Schema [ISO 03] and Simple Feature Access [ISO 00].

The Spatial Schema and Simple Feature Access standards govern the use of Egenhofer's nine-intersection model, as well as Clementini's extensions. For example, the Simple Feature Access standard requires the use of an intersection pattern to express the nine intersection values for a given relationship. The relationship represented by the matrix in Figure 9.2 can be expressed as a pattern by stringing the values together from left to right and top to bottom to produce "TTFFTTTT".

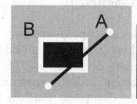

	Interior *(b)*	Boundary *(b)*	Exterior *(b)*
Interior *(a)*	T	T	T
Boundary *(a)*	F	F	T
Exterior *(a)*	T	T	T

Figure 9.3. *Example of the intersection matrix using only true or false to describe the presence of intersections (T or F)*

whereas the same relationship represented by the matrix in Figure 9.4 can be expressed as the pattern *101FF0212*.

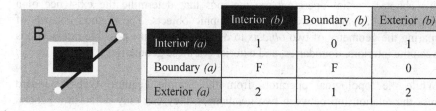

	Interior *(b)*	Boundary *(b)*	Exterior *(b)*
Interior *(a)*	1	0	1
Boundary *(a)*	F	F	0
Exterior *(a)*	2	1	2

Figure 9.4. *Example of an intersection matrix returning the geometry for cases involving an intersection (0, 1 or 2)*

The Simple Feature Access standard uses the nine-intersection model and Clementini's extensions to define eight topological-relationship predicates: equals, disjoint, intersects, touches, crosses, within, contains, and overlaps. Each predicate is defined in Table 9.3.

Equals	Geometry that is "spatially equal" to another geometry.
Disjoint	Geometry that is "spatially disjoint" from another geometry.
Intersects	Geometry that "spatially intersects" another geometry of smaller dimension.
Touches	Geometry that "spatially touches" another geometry.
Crosses	Geometry that "spatially crosses" another geometry.
Within	Geometry that is "spatially within" another geometry.
Contains	Geometry that "spatially contains" another geometry.
Overlaps	Geometry that "spatially overlaps" another geometry.

Table 9.3. *Topological-relationship predicates according to ISO*

9.4. Definitions and concepts: the components of integrity constraints

This section defines the various concepts used to describe spatial integrity constraints. We explain how CTI uses the topological-relationship predicates found in international standards in geomatics, as well as the extensions developed by CTI. These extensions have proven necessary to the development of the Canadian government's entirely new geospatial database called the Geospatial Database (GDB) of Canada. We also describe the cardinality associated with predicates as well as buffers and their parameters.

9.4.1. Spatial operators

Spatial operators are functions that have at least one spatial parameter [ISO 03]. In this chapter, spatial operators are methods that determine the existence of a topological relationship between two geographic objects. The method consists of comparing the geometry of two *objects* to determine the nine possible intersections between the interiors, boundaries, and exteriors of these geometries.

While the topological operators from the Simple Feature Access standard express the relationship between two geometries, they can only do so in a general manner. For example, the left portion of Figure 9.5 depicts three different cases in which an object is within another. In the GDB, extensions are required to express this relationship in these special cases.

Three additional notions have been defined in order to extend the predicates in the Simple Feature Access standard: *tangent*, *borders*, and *strict*. These three notions are used with the predicates from the standard, making them more specific. The right-hand portion of Figure 9.5 shows the same situations with the three extensions used to refine the predicate "within".

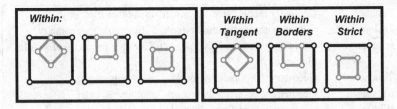

Figure 9.5. *Situations responding to the predicate within (on the left) and refinement of the predicate by adding the extensions tangent, borders, and strict (on the right)*

9.4.1.1. *The extension tangent*

The extension tangent specifies that the boundaries of the geometric objects must intersect and that the intersection dimension is 0 (□). It can be combined with the predicates touches, within, and contains, and applies to the geometries such as ☑/☑, ☑/▣, ▣/☑, and ▣/▣:

Touches ∧ Tangent = Touches ∧ (dim(B(a) ∩ B(b)) = 0)

Within ∧ Tangent = Within ∧ (dim(B(a) ∩ B(b)) = 0)

Contains ∧ Tangent = Contains ∧ (dim(B(a) ∩ B(b)) = 0).

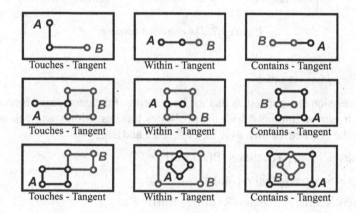

Figure 9.6. *The extension tangent*

9.4.1.2. *The extension borders*

The extension borders specifies that, when a polygon intersects another polygon, there must be an intersection between polygon boundaries and the intersection dimension is 1 (☑).

It can be combined with the predicates touches, within, and contains, and applies to the geometries such as ▣/▣:

Case a ∈ ▣, b ∈ ▣:

Touches ∧ Borders = Touches ∧ (dim(B(a) ∩ B(b)) = 1)

Within ∧ Borders = Within ∧ (dim(B(a) ∩ B(b)) = 1)

Contains ∧ Borders = Contains ∧ (dim(B(a) ∩ B(b)) = 1).

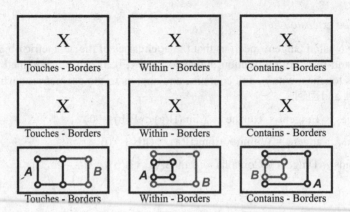

Figure 9.7. *The extension borders*

9.4.1.3. *The extension strict*

The extension strict specifies that the boundaries of the geometric objects cannot intersect. It can be combined with the predicates touches, within, and contains, and applies to the geometries such as ☑/☑, ☑/☒, ☒/☑, and ☒/☒:

Touches ∧ Strict = Touches ∧ $(B(a) \cap B(b) = \varnothing)$

Within ∧ Strict = Within ∧ $(B(a) \cap B(b) = \varnothing)$

Contains ∧ Strict = Contains ∧ $(B(a) \cap B(b) = \varnothing)$.

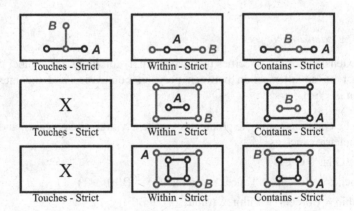

Figure 9.8. *The extension strict*

9.4.2. *Cardinality associated with predicates*

Cardinality is an additional component that specifies how spatial operators shall be used. By definition, a spatial operator compares a pair of objects and returns a binary response (true or false) to the test carried out. Cardinality specifies the number of objects with which an object can have a relationship. For example, a linear segment representing a ferry route must have a relationship with a road segment (touches). This ensures that no ferry is isolated from road infrastructure. In reality, however, one could expect a ferry crossing a waterbody to have a relationship with two road segments, namely, one at each end of the ferry route. This type of validation is an integral part of the spatial integrity constraints between geographic objects. Cardinality is represented by a pair of values that specify the minimum and maximum number of geographic objects with which an object can have a relationship. For example, in the case of the linear segment representing a ferry route, the spatial constraint (Ferry Route Touches Road) has a cardinality of (2,n), which specifies that a minimum of two segments of road must touch the ferry route segment in order to comply with the constraint (Figure 9.9).

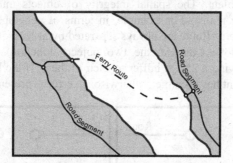

Figure 9.9. *Example of a ferry route segment touching two road segments (one at each end)*

9.4.3. *Buffer*

A buffer is also used with special operators in dealing with two specific problems: coordinate precision and the proximity of objects in geographic reality.

On the one hand, the computer algorithms underlying spatial operators normally require that two points or vertices with identical coordinates be considered as equal. In reality, many of the software applications that process geometric data use different levels of precision (that is, number of meaningful decimal places) and sometimes different data types (integer versus real) to record coordinates (integer versus real). This makes it difficult to use special operators, especially when processing data from heterogeneous sources. Indeed, special operators indicate errors related to spatial integrity constraints, whereas the spatial position of objects falls within a given tolerance and can be automatically corrected, if necessary. Certain commercial software applications work around this problem by using a very fine grid and associating object coordinates to grid intersections. This approach rounds coordinates from each data geographic source to a common reference for greater accuracy in identifying errors with respect to spatial integrity constraints. This means that spatial integrity constraints are validated with a virtual version of data with rounded coordinates, not with the original data. To meet the needs of the GDB, CTI has opted for an approach that defines a buffer around objects that is taken into account when validating spatial integrity constraints. Figure 9.10 illustrates a case in which two objects (Junction – Dead End (that is, road or street without issue)) shall be disjoint by 1 m or more from one another. These two objects are compared after applying a 0.5-m buffer to each. If the buffers are disjoint, the integrity constraint has been met; otherwise, an error is present (Figure 9.10).

On the other hand, object proximity in geographic reality represents another spatial-integrity problem. The spatial integrity of objects must conform to the observed geographic reality. For example, in terms of geographic reality, the two objects Junction – Ferry Route are always separated by at least 500 m (Figure 9.11). Similarly to the above example, the two objects Junction – Ferry Route are compared after applying a 250-m buffer to each of them. If the buffers are disjoint, the integrity constraint has been met; otherwise, an error is present.

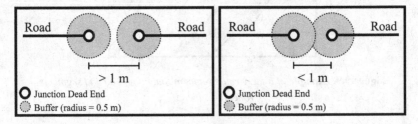

Figure 9.10. *Example of two objects (Junction – Dead End) disjoint by more than 1 m (on the left) and less than 1 m (on the right)*

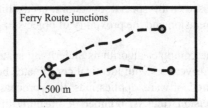

Figure 9.11. *Example of two objects (Junction – Ferry Route) disjoint by more than 500 m*

9.5. Documentation and use of integrity constraints

When developing geographic databases, geographic object cases, their descriptive properties, their value domains, and their geometry, the relationships between classes, and so on should be documented in a data repertory. Preferably, it should be a UML model and data dictionary, or the equivalent, that strengthens international standards in geomatics [ISO 02, ISO 04]. The data repertory must also include documentation on spatial integrity constraints. The next section deals with documenting spatial integrity constraints as described above.

9.5.1. Documenting spatial integrity constraints

The spatial integrity constraints that apply to classes in a geographic database should be documented in a geographic repertory. A spatial-integrity constraint comprises geographic object classes and any related specific descriptive properties (if applicable), the name of the topological predicate, the cardinality, and the buffer size, if appropriate. Figure 9.12 provides an example of documentation of spatial integrity constraints from the GDB geographic repertory. The spatial integrity constraints are presented according to object class (for example, road segment). Each object class contains groups with one or more constraints. All of the constraint groups must be complied with, which means that the operator AND will appear between all of the groups (for example, B1 ∧ B2 ∧ B3 ∧ B4). At least one constraint must be complied with within each constraint group. This means that the operator OR appears between all of the constraints (for example, C1 ∨ C2 ∨ C3).

Road Segment

Road Segment Does Not Overlap (10 m) Road Segment [-,-]

Road Segment Touches-Tangent to Road Segment [1,2]

Road Segment (*Structure Type* = *'None'*) Disjoint to Road Segment (*Structure Type* = *'None'*) [-,-]
Road Segment (*Structure Type* = *'None'*) Touches-Tangent to Road Segment (*Structure Type* = *'None'*) [1,2]

Road Segment (*Structure Type* <> *'None'*) Crosses Road Segment [1,n]
Road Segment (*Structure Type* <> *'None'*) Disjoint to Road Segment [1,n]
Road Segment (*Structure Type* <> *'None'*) Touches-Tangent to Road Segment [1,n]

Figure 9.12. *Spatial integrity constraints from the GDB repertory*

Since these integrity constraints are stored in a computerized repertory, they can be easily extracted and presented as a table, as illustrated in Figure 9.12, or in another format, such as an XML (Extensible Markup Language) document.

9.5.2. *Validation of integrity constraints (inconsistency)*

In order to ensure their integrity, geographic data are validated with respect to the spatial-integrity constraints as a whole. Data validation begins by revealing a series of inconsistencies with respect to integrity constraints. These inconsistencies are represented by point, linear, surficial, or multiple geometric objects and are spatially referenced. One or more inconsistencies may be found for each of the constraints.

After inconsistencies have been detected, the data are corrected, either automatically or interactively depending on the nature of the inconsistency. For example, if two point objects from the same class are inappropriately superimposed, automatic processing can eliminate one of the objects. Other instances require human intervention.

In certain situations, additional information is required in order to correct the inconsistency. When no additional information is available, the inconsistencies between objects are described, located, and preserved in the dataset just as an object would be. In this way, the producer is aware of the inconsistencies in a dataset and they serve as warnings to users. This situation can arise when data from outside sources, such as a provincial or municipal cartographic department that uses different specifications for capturing data, are integrated into an existing geographic database. To illustrate, attempting to integrate vegetation polygons representing coverage of at least 35% of trees over 2 m in height with vegetation polygons representing coverage of 50% or more of trees over 2 m in height can give rise to instances of misclosure (Figure 9.13).

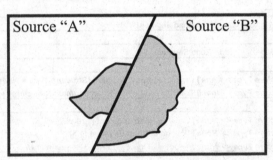

Figure 9.13. *Vegetation polygons from two different sources with different object class definitions*

9.6. Production and validation of geographic data

The description and validation of spatial integrity constraints discussed above are part of a larger process in effect at NRCan. This process involves a variety of phases in which geographic data are produced through contracting out to private-sector firms. The production process consists of the following phases (Figure 9.14):

– Planning: selection of the classes and portion of the territory for processing or updating.

– Tendering: posting of the specifications for the work to be carried out.

– Awarding contracts: awarding of the contracts to the selected firms.

– Automatic validation: all of the computer operations for validating data structure, such as encoding, attribute domain values, object minimum size, and spatial integrity constraints.

– Interactive validation: all the human operations that complement automatic validation. Certain inconsistencies detected during the previous phase are confirmed or invalidated at this point.

– Delivery: the data are entered into the GDB.

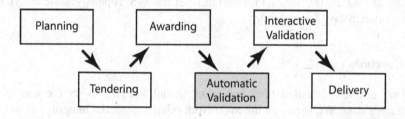

Figure 9.14. *Geographic data production process used at NRCan*

9.6.1. *Validating the spatial integrity of geographic data*

A certain number of validations need to be made before spatial integrity constraints can be validated. Spatial primitives (point, line, surface) must be respected. The mandatory attributes must be present and their values must fall within their respective value domains. This is of particular importance when certain spatial integrity constraints are verified for only a data subset based on an attribute value for an object class (for example, a specific class – road, loose surface). Consequently, automatic validation contains a well-ordered sequence of routines, each of which validates a specific aspect of data structure:

– data coding;

– segmentation at the dataset neatline;

– oversegmentation;

– missing segmentation;

– filtering of geometric objects;

– spatial integrity constraints (spatial relations, cardinality, buffer).

9.6.2. *Available tools*

Many designers of GIS tools (both commercial and open-source applications) include spatial analysis operators in their software suites. While specific to each designer, these operators are being converted to conform to ISO 19100.

Since software and computers change rapidly, as a general rule we have not deemed it necessary to present an inventory of software that offer spatial analysis operators and their characteristics. Nevertheless, we considered it important to mention the JUMP-Project [JUM 04], which designed an open-source Application Programming Interface (API) for validating all the spatial integrity constraints defined in ISO 19100. This API, referred to as the JTS Topology Suite (JTS), has been written in Java for portability.

9.7. Conclusion

This chapter presented how important spatial integrity is to the quality of geographic data. We reviewed the main work related to spatial integrity, as well as the associated international standards in geomatics. We then presented CTI's approach to describing, documenting, and validating spatial integrity constraints.

In conclusion, the definition of spatial integrity constraints between object classes and the validation of geographic data add value to the overall quality of geographic data. The validation of these constraints during data production and updating minimizes the quantity of errors and ensures better integrated data for users.

Moreover, a single database comprises data from various sources, the object classes must comply with a common semantic definition to minimize inconsistencies when fusing data.

Moreover, the use of international standards makes it possible to standardize spatial integrity constraints, so that tools that produce compatible results can be developed.

Lastly, spatial integrity constraints must be meticulously defined if they are to be functional and useful tools. Certain aspects, such as reciprocity, will significantly influence the processing burden of validating constraints. For example, an initial relationship may state that no road can cross a waterbody and a second one the opposite, namely, that no waterbody may cross a road. This situation entails two operations, which doubles the processing required to validate a single constraint. It is important to properly structure and use the geographic repertory that contains the spatial integrity constraints and to carefully manage constraints.

9.8. References

[BED 89] BÉDARD Y. and PAQUETTE F., "Extending entity/relationship formalism for spatial information systems", *9th International Symposium on Automation in Cartography (AUTO-CARTO 9)*, 2–7 April 1989, American Congress on Surveying and Mapping, Baltimore, USA, p 818–27.

[BED 99] BÉDARD Y., "Visual modelling of spatial database: towards spatial PVL and UML", *Geomatica*, 1999, vol 53, number 2, p 169–86.

[BOR 02] BORGES K.A.V., DAVIS C.A. and LAENDER A.H.F., "Integrity constraints in spatial databases", in *Database Integrity: Challenges and Solutions*, J.H. Doorn and L.C. Rivero (eds), 2002, Hershey, Idea Group Publishing, p 144–71.

[CLE 95] CLEMENTINI E. and Di FELICE P., "A comparison of methods for representing topological relationships", *Information Sciences*, 1995, vol 3, p 149–78.

[CLE 96] CLEMENTINI E. and DI FELICE P., "A model for representing topological relationships between complex geometric features in spatial databases", *Information Sciences*, 1996, vol 90, number 1–4, p 121–36.

[COC 97] COCKCROFT S., "A taxonomy of spatial data integrity constraints" *GeoInformatica*, 1997, vol 1, number 4, p 327–43.

[COC 04] COCKCROFT S., "The design and implementation of a repository for the management of spatial data integrity constraints", *GeoInformatica*, 2004, vol 8, number 1, p 49–69.

[EGE 91] EGENHOFER M.J. and FRANZOSA R.D., "Point-set topological spatial relations", *International Journal of Geographical Information Systems*, 1991, vol 5, number 2, p 161–74.

[EGE 93] EGENHOFER M.J., "A model for detailed binary topological relationships", *Geomatica*, 1993, vol 47, number 3–4, p 261–73.

[EGE 94a] EGENHOFER M.J., CLEMENTINI E. and DI FELICE P., "Topological relations between regions with holes", *International Journal of Geographical Information Systems*, 1994, vol 8, number 2, p 129–42.

[EGE 94b] EGENHOFER M.J., MARK D.M. and HERRING J.R., *The 9-Intersection: Formalism and Its Use for Natural-Language Spatial Predicates*, 1994, Report 94–1, University of California, NCGIA, Santa Barbara, USA.

[EGE 95] EGENHOFER M.J. and FRANZOSA R.D., "On the equivalence of topological relations", *International Journal of Geographic Information Systems*, 1995, vol 9, number 2, p 133–52.

[EGE 97] EGENHOFER M.J., "Spatial relations: models and inferences", *Proceedings of the 5th International Symposium on Spatial Databases (SSD'97)*, Tutorial 2, 15–18 July 1997, Berlin, Germany, p 83.

[HAD 92] HADZILACOS T. and TRYFONA N., "A model for expressing topological integrity constraints in geographical databases", in A. Frank, I. Campari and U. Fomentini (eds), *Lecture Notes in Computer Science 639*, 1992, Springer-Verlag, New York, p 348–67.

[ISO 00] ISO/TC 211, ISO/DIS 19125-1 *Geographic Information – Simple Feature Access – Part 1: Common Architecture* (International Organization for Standardization), 2000.

[ISO 02] ISO/TC 211, ISO/DIS 19109 *Geographic Information – Rules for Application Schema* (International Organization for Standardization), 2002.

[ISO 03] ISO/TC 211, ISO 19107:2003 *Geographic Information – Spatial Schema* (International Organization for Standardization), 2003.

[ISO 04] ISO/TC 211, ISO/FDIS 19110 *Geographic Information – Feature Cataloguing Methodology* (International Organization for Standardization), 2004.

[JUM 04] The JUMP-Project, http://jump-project.org/, 2004.

[MOS 04] MOSTAFAVI M.A., EDWARDS G. and JEANSOULIN R., "Ontology-based method for quality assessment of spatial data bases", *Proceedings of the Third International Symposium on Spatial Data Quality* (ISSDQ '04), 2004 GeoInfo Series, vol 28a, Bruck an der Leitha, Austria, p 49–66.

[NOR 99] NORMAND P., Modélisation des contraintes d'intégrité spatiales: Théorie et exemples d'application, 1999, Master's thesis, Department of Geomatics, Laval University, Quebec.

[PAR 97] PARENT C., SPACCAPIETRA S., ZIMANYI E., DONINI P., PLAZANET C., VANGENOT C., ROGNON N. and CRAUSAZ P.A., "MADS, modèle conceptuel spatio-temporel", *Revue Internationale de Géomatique*, 1997, vol 7, number 3–4, p 317–52.

[RUM 91] RUMBAUGH J.R., BLAHA M.R., PREMERLANI W., EDDY F. and LORENSEN W., *Object-Oriented Modeling and Design*, 1991, Englewood Cliffs, NJ, Prentice-Hall.

[UBE 97a] UBEDA T., Contrôle de la qualité spatiale des bases de données géographiques: cohérence topologique et corrections d'erreurs, 1997, Doctoral thesis, Institut of Applied Sciences in Lyon, Lyon, p 205.

[UBE 97b] UBEDA T. and EGENHOFER M.J., "Topological Error Correcting in GIS", *Proceedings of the 5th International Symposium on Spatial Databases* (SSD '97), Berlin, Germany, 15–18 July 1997, *Lecture Notes in Computer Science*, vol 1262, Springer-Verlag, p 283–97.

Chapter 10

Quality Components, Standards, and Metadata

10.1. Introduction

Several years ago, databases stopped being merely simple collections of information stored in a structured format and became what they are today: indissociable from information systems (IS) that use their data and of which they are part.

Such information systems form the core of various applications, both at the final level (management, systems for helping decision-making, etc.) as well as at the level of end users (banks, local governments, large organizations, etc.).

In such a context, it is essential to understand what data are, and to control its quality. This necessitates the active involvement of designers of IS and the producers of the underlying data to ensure that the data fulfills the needs of the future users.

Existing geographic databases often contain errors due to the acquisition source (measuring instruments), data-input processes, and information processing. In addition, the evolution in features of geographic information systems (GIS), as well as the emergence of the Internet, has caused a shift in how information systems and their underlying data are used; shared information that is available online can be "diverted" from its primary use. Mainly due to its high acquisition costs, spatial data tends to have a long life – which leads to it being used for purposes that were not originally foreseen. Originally acquired to allow cartographic plotting (which could

Chapter written by Sylvie SERVIGNE, Nicolas LESAGE and Thérèse LIBOUREL.

accommodate errors that were not visible at the plotting scale), entire datasets are now being used in the field of spatial analysis which uses methods that range from interpolation to simulation for the purpose of helping in decision-making. Limitations, in terms of quality, of such data are more significant in this type of processing (topological consistency, accuracy of attribute values, etc.) and it becomes imperative to define quality standards and strategies to improve this quality so that the life of currently existing batches of data and datasets can be extended. Moreover, if accuracy and reliability have long been the parameters of quality for qualifying geodetic networks, the quality of today's spatial or spatio-temporal databases is more difficult to define because of the complexity of spatial attributes: dimensions of definition of managed objects (1D, 2D, 3D geometric descriptions), spatial relationships between the objects (topology), potential richness of non-spatial attributes, etc.

The design of IS and databases should include, in its own right, the data quality. Thus, the quality should be specified and processed for improving and monitoring it implemented [SER 00]; some data changes rapidly (notably in the urban environment) and the data quality should also be ensured over the long term.

This integration of data and data quality is most often implemented by using metadata ("data about data" in its first meaning). Metadata allow the documentation, as precisely as possible, of data, facilitating its sharing and distribution with the overall goal of simplifying its integration and reuse. The emergence of the digital document has led to the phenomenon of annotation that is well known to librarians. The proposals of the Dublin Core and W3C (World Wide Web Consortium) attest to the use of annotation. However spatial information, due to its particularities, requires complex and voluminous metadata to be stored and organized for geographic information, and this complexity and size becomes a major hindrance to its wider use. It is therefore imperative that efficient and well-conceived standards exist and take into account data quality in the appropriate measure. As an example, the information on quality should ensure the reliability of processes based on the data, as well as the system's ability to fulfill expected functions (that is, suitability for requirements as expressed in the specifications). These two complementary notions are found in the definition of quality put forward by the International Organization for Standardization [ISO 94]: "Set of properties and characteristics of a product or a service which confers upon it the ability to satisfy expressed or implicit requirements." Finally, in the context of geographic information, it is necessary to keep in mind the different points of view of different users of geographic data, that is, the data producers and data users. In fact, data producers or suppliers want to adhere to quality standards because it confers certification on batches of data that they produce or sell. Users, on their side, would like to have data the quality of which is appropriate to their needs and thus to their applications.

This chapter thus takes up the concepts of quality introduced in standardization approaches. It will describe their definition and how they have been incorporated within metadata standards dedicated to spatialized information.

10.2. Concepts of quality

For a long time, the description of quality has been reduced to a problem of the accuracy of stored information (see Chapter 1). During the analogue age, as far as geographic data was concerned, accuracy almost exclusively concerned the position of represented objects, ignoring problems linked to their shape, representation, semantic quality or consistency. The advent of the digital age saw work on standardization, which started in the 1980s, leading up to a consensus on the definition of quality components.

The terminology surrounding spatial data quality is subject to numerous variations, and different terms are sometimes used to describe the same concept.

10.2.1. *Quality reference bases*

To ascertain a dataset's quality, it is necessary to have reference bases that will serve as a basis of comparison of the datasets under consideration. Two concepts, "nominal ground" and "universe of discourse", can constitute possible definitions of the reference base:

– Nominal ground: A literal, though inexact, translation of the French term, *"terrain nominal"*. It has a number of definitions, among which that of the Institut Géographique National (IGN), adopted by the CEN [CEN 98]: "the real world as viewed through the specifications used when inputting a batch of geographic data." For a shorter version, we can consider the definition of [VAU 97], "that which should have been entered", as wholly satisfactory to describe the notion of nominal ground and corresponding better to the English definition of "universe of discourse", which clearly separates the producer and user aspects;

– Universe of discourse: abstractions of the real world incorporating both complementary aspects:

- of data-production specifications;

- of users' requirements, etc.

10.2.2. *Quality criteria*

Criteria called quantitative embody a quantitative expression of quality (for example, spatial accuracy = 12 m) and are also called "quality parameters'. Criteria called qualitative (truth in labeling) provide a qualitative expression of quality (example: lineage).

In 1987, the National Committee on Digital Cartographic Data Standards (NCDCDS) ([MOE 87]) proposed the definition that describes spatial data quality by breaking it down into five criteria (one qualitative and four quantitative): lineage, geometric/positional accuracy, semantic/attributes accuracy, completeness, and logical consistency. In 1991, the executive committee of the International Cartographic Association (ICA) established a commission on data quality [GUP 95]. This commission had an aim to develop, document, and publish criteria and methods for the evaluation of digital cartographic datasets. It identified three parts in the specification and use of information on the quality of spatial data:

– the definition of elements of spatial quality;

– the establishment of metrics to measure elements of spatial quality;

– the presentation (communication) of data quality.

In 1995, to the five quality criteria defined by the NCDCDS, the commission added two new parameters: "temporal accuracy" and "semantic consistency". In 1997, the IGN [DAV 97] introduced "specific quality" to help overcome potential lacunae not covered by the previous criteria.

10.2.2.1. *Qualitative criterion*

The qualitative criterion retained is designated by the term "lineage".

This criterion provides the material origin of the data and the methods used, as well as all subsequent transformations undergone by the data, to arrive at the final data. In other words, the lineage describes the acquisition procedures and methods of deriving and transforming data.

However, the objective of the lineage component can be interpreted differently by the data producer and the data user:

– the producer wants to ensure that standards are maintained;

– the user wants to know the origin of the acquired data so that he can be sure that it fulfills his requirements.

10.2.2.2. *Quantitative criterion*

The quantitative criterion or quality parameters are:

– *Geometric accuracy (or positional accuracy, spatial accuracy)*. It gives the level of conformity of data with respect to the nominal ground from the point of view of the difference of the respective positions in these two views. It thus defines the deviation in the values of the respective positions between the data and the nominal ground.

– *Semantic accuracy*[1] *(or accuracy of non-spatial attributes)*. This criterion provides information on the difference between the values of non-spatial attributes and their real value and thus gives us the deviations of measurements of qualitative attributes or quantitative attributes (classification).

– *Completeness*. It can be applied to the level of the model, the data or even objects and attributes. Data completeness helps us to detect errors of omission (abnormal absence) or commission (abnormal presence) of certain objects. Model completeness, on the other hand, expresses suitability of the provided representation for users" requirements.

– *Logical consistency*. It has the goal of describing the faithfulness of relationships encoded in the database's structure with respect to all the constraints caused by data-input specifications. In other words, it describes the correspondences of the dataset with the characteristics of the structure of the model used (respecting specified integrity constraints).

– *Temporal accuracy*. It provides information on the temporal aspect of data: management of data observation dates (origin), of types and frequency of updates, and the data's validity period. It could be essential to have this information, especially when evaluating the suitability for the requirements of a particular user.

– *Semantic consistency*. It indicates the relevance of the significance of objects with respect to the selected model; it describes the number of objects, of relationships, and attributes correctly encoded with respect to the set of rules and specifications.

– *Specific quality*. A quality parameter (thus quantitative) which expresses quality-related information that is not foreseen by the previous criteria. Thus, IGN [DAV 97] introduced the concept of "timeliness" which helps determine the suitability for requirements by translating the offset between the produced dataset and the nominal ground to a later instant.

10.2.3. *Expression of the quality*

Quality is expressed with the help of indicators, elements, and measurements; their definitions are as follows:

– *Quality indicator*. Set of quality measurements indicating the performance of a quality parameter for an entire batch of geographic data.

---

1 The semantic qualifier was initially associated with non-spatial attributes and, even though this nomenclature can be debated, we retain the qualifier.

– *Quality element*. Set of quality measurements indicating the performance of a quality parameter for all or part of a batch of geographic data.

– *Quality measurement*. Definition of a specific test to apply to geographic data, including algorithms, and the type of value or set of values that result.

10.2.4. *Precision and accuracy*

There is a fundamental difference between the two concepts: precision indicates the *resolution* with which one can measure a phenomenon with a particular instrument or method (see Figure 10.1), as well as the ability to obtain the same value by repeating a given measurement. In the GIS domain, precision varies most often with the cartographic scale used. A rule of thumb is that a precision is acceptable if it causes an error on the map in the order of 1/10th of a millimeter (which at 1:1,000 represents an error of 10 cm and at 1:500,000, an error of 50 m). On the other hand, accuracy bears on the notion of truth (the center of the target in Figure 10.1), and of exact data representing faithfully the real phenomenon that it is attempting to represent. Inaccuracy arises from, among other things, measurement errors and can be linked to systematic methodological problems, themselves caused by the imperfect nature of the method used to acquire the data, and by use of unsuitable digital processing procedures (for example, a numeric range that is too narrow in a series of complex calculations with automatic truncation at each step of the process). These systematic errors should, as far as possible, be listed in the lineage elements (see section 10.3.1), even if their effects are also felt, for example, in the domain of the geometric precision.

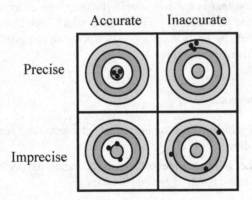

Figure 10.1. *Comparison: accuracy and precision*

10.2.5. *Appraisal and use of quality*

A very important concept for the appraisal of quality by the user is that of suitability for requirements or "fitness for use" (often called "external quality"). It represents the potential – admittedly subjective – of the data to fulfill specific requirements of the user (see Chapter 13 for an example of a method for evaluating fitness for use). This is a difficult criterion to evaluate using the quality criteria defined above. Nevertheless, it is essential to do so because it allows a potential user to determine whether a particular dataset can fulfill the purpose he expects it to. Tests of deviation, appropriate to the target application, will have to be available or complementary annotation by the user will have to be authorized (based on the meta-quality and the user's specific expertise).

10.2.6. *Meta-quality*

The evaluation of the quality, using any parameter, allows us to represent the corresponding performance of the dataset with respect to the considered quality element. It is essential to supply, at the same time as the result of the evaluation, a set of indications that allows one to qualify this information. We are now talking of quality of quality, and use the term "meta-quality" to describe it. The most important of these indications are the date of processing (temporal aspect), the evaluation method used (tested, calculated or estimated), and the population on which it was applied:

– The processing date could be *ad hoc* (in the case of a quality audit conducted at regular or irregular intervals) or could be continuous, as in the case of systems for which mechanisms exist to ensure integrity of some data aspects (triggers, etc.). The processing date then corresponds to the date the quality report was created.

– The methods used could be more, or less, reliable (use of a threshold, quality of algorithms used, propagation methods – statistics).

– Finally, the population will vary depending on the method: from the entirety of the data for a general audit, to different types of sampling involving a variable number of elements. Partitions can also be used, either temporal (evaluation of the quality of entities input in the last two months, or of those that are three to five years old) or geographic (processing of a specific administrative area, for example). These two types of partitions can, of course, also co-exist within the same process.

CEN [CEN 98] has identified three main elements of meta-quality. These are confidence, homogeneity and reliability:

– *Confidence*: "a meta-quality element that describes the accuracy of quality information." Confidence originates primarily from the method used and of its reliability, as well, to a lesser extent, from the concerned population.

– *Homogeneity*: "textual and qualitative description of the expected or tested uniformity of quality parameters in a batch of geographic data." In fact, a dataset can be the result of a single acquisition process or result from a combination of various acquisition techniques (aerial photos, digitization of paper maps, GPS, theodolites, etc.). The homogeneity depends mainly on the population that was the basis of the evaluation. In the case of a general process, it cannot be evaluated because the result is global. Homogeneity is thus only relevant when several segments were used and their evaluation results (derived using the same methods) compared. These tests are often conducted when data has been input by different operators, depending on the zone or the acquisition date.

– *Reliability*: "a meta-quality element describing the probability that a given sampling of a batch of geographic data, when used for quality evaluation purposes, is representative of the entire data batch." A statistical method based on sampling could be considered as reliable as a global method when all the geographic zones and concerned time periods are covered and the population is sufficiently large.

10.3. Detailed description of quality criteria

10.3.1. *Lineage*

[CLA 95] identifies the information necessary for reconstructing the history of a dataset and to deduce therefrom its potential usage (processing methods and tools for a particular requirement):

– The data source (the organization's reputation, if not quantifiable, should also be taken into account), origin, reference domain (geology, etc.), characteristics of spatial data, coordinate and projection systems, and associated corrections and calibrations.

– Acquisition, compilation and derivation: fundamental hypotheses of observation, calibration, and corrections. Then the geo-referencing or application to a particular domain – taking an arbitrary 0 altitude, for example – followed by the description of methods used to interpret, interpolate or aggregate data, at the level of the structure or the format used.

– Data conversion: definition of processes, such as, for example, the stages in the vectorization of raster data.

– Dates of different stages of processing.

– Transformations or analyses: transformation of coordinates, generalization, translation, reclassification, all defined, as far as possible, in precise mathematical terms. All parameters used should be clearly defined, since these transformations can have profound effects on the produced data.

At the normalization level, importance is often accorded to the data structure rather than to its semantics. It is possible that the real nature of the information on the lineage is not sufficiently "closed" so as to be able to be represented in a standardized manner (the number of possible and successive processes perhaps ruling it out). In any case, lineage information is often provided in the form of running text that describes the parameters listed above.

The collection of this information can prove to be an onerous and difficult task, especially when it concerns data originating from different acquisition processes, and having undergone numerous transformations. It is, however, in this type of case that it is most useful, indeed indispensable.

10.3.2. *Positional accuracy or geometric accuracy*

Positional accuracy is generally divided into absolute accuracy and relative accuracy. It can also be differentiated between planimetric accuracy and altimetric accuracy (for 3-dimensional data). Altimetric accuracy often comes down to a problem of semantic accuracy (see next section), since the altitude of points is often stored in the form of an alpha-numeric attribute.

The position of objects in the database is a set of cardinal values that allow the objects to be positioned in three-dimensional Cartesian or polar coordinates. For example [AZO 00]: field mapping (X, Y, Z), GPS position (latitude, longitude, altitude), digitization (Y, X). The only way to measure positional accuracy is, therefore, to compare the dataset, either with another dataset of better quality (and following the same specifications), also called "control" or "reference" data [DAV 97], or with data derived from surveys and samplings (for example, with a GPS sensor). Geometric accuracy, or accuracy of the coordinates, directly depends on the acquisition methods and processing of measurements. For example, the positional and altitudinal accuracy of contour lines depends on the accuracy of measurement of the points used to determine the contours and of the interpolation algorithms used. It specifies root mean square errors (RMSE) in planimetry and altimetry in the points' coordinates, possibly even their mean error ellipse.

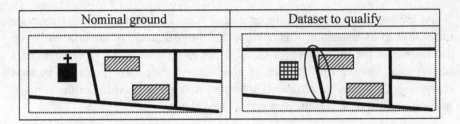

Figure 10.2. *Example of positional inaccuracy*

10.3.3. *Attribute accuracy or semantic accuracy*

An attribute can be the result of a measurement or interpretation, originate from direct human observation (such as the names of roads or lakes), or even from a historical or political census [UBE 97].

In the same manner as for geometric accuracy, the semantic accuracy is defined as the difference between a measurement and another comparable measurement known to be more accurate. This is a relative definition because it relates to the accuracy of the objects being compared. It also requires the knowledge of more accurate data, namely the nominal ground. As this does not really physically exist, reference data are used instead of the nominal ground.

All types of attributes are subject to uncertainty because of defects in measuring instruments or data-acquisition procedures, or historical uncertainty that can afflict names. These uncertainties can be of different types depending on whether the attribute applies to a single location (attribute that is difficult to measure or valid only at a certain scale) or on a set of points (attributes are often calculated as averages or aggregations of values in the area under consideration).

To help evaluate semantic accuracy, a classification according to a scale of measurement was created for the specific requirements of spatialized information [GOO 95]. This classification applies to different types of simple attributes, that is, attributes that are qualitative (names, classes used to characterize data) and quantitative (measurements, enumerations, analysis results, etc.) and introduces:

– *Nominal* scales (used to classify some characteristics, and though often numbered, not representing numerical values), such as residential, commercial or industrial zoning.

– *Ordinal* scales (to classify and sort), such as the soil richness: poor, medium or rich.

– *Interval* scales (when the system uses a relative zero – only measured differences make sense, as for temperature expressed in degrees Celsius – the difference in temperature is the same between 10 and 20°C as it is between 20 and 30°C; but 40°C is not the double of 20°C since the zero is arbitrary).

– *Ratio* scales (if the ratios between measurements make sense, as is the case with temperatures in Kelvin for which 200 K = 2 × 100 K, since this scale is based on an absolute zero).

The first two scales can define both qualitative and quantitative attributes, whereas the latter two can only define quantitative values.

For attributes with cardinal values for example (interval or ratio), standard deviation can be used or, if necessary, an estimate of this standard deviation (height of trees estimated at ± 10%). For attributes with ordinal values, it becomes necessary to qualify the accuracy of the classification of objects when, for example, there is a possibility of confusion between object classes (for example, are the vegetation zones identified on an aerial photo not, in fact, constructed zones?). As for nominal values, a descriptive entry could be used to alert the user to the accuracy of the text. For Azouzi, for example [AZO 00], since the designation is one of the attributes of a building, a qualifier of this attribute allows the user to be aware of the difficulties encountered during the assigning of the designation. By their very nature, errors linked to different types of attributes follow different statistics.

The determination of the semantic accuracy is sometimes similar to completeness if one considers that a difference in conceptual modeling can transform an attribute to a class or vice versa. Similarly, the geometric accuracy becomes a sort of semantic accuracy when we treat the location of objects as a specific attribute of entities [GOO 95].

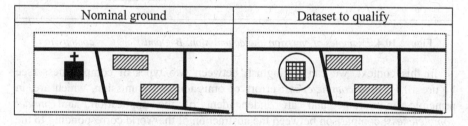

Figure 10.3. *Example of semantic inaccuracy*

10.3.4. *Completeness*

"Completeness is an attribute that describes the relationships between objects represented in a dataset and is an abstraction of the same set of objects in the real world" [MOE 87]. Evaluating objects of the database with all the objects of the universe of discourse requires, therefore, that a formal description of both these sets be available.

Thus, depending on the domain under consideration, the completeness of a database (or a map) can be suitable for a specific task, but not for another. One has therefore to relate the data quality with the fitness for use. The concept of "fitness for requirements" or "fitness for use" comes into its own when data completeness has to be measured. In fact, if the information on data quality is, in principle, supplied by the producer of the dataset, the fitness for use, on the other hand, is only estimated at the time of evaluation of the *use* of the dataset (principle of "truth in labeling"). In the useful lifetime of a dataset, the quality (considered in a general manner and not only for completeness) will be evaluated only once, whereas a fitness-for-use evaluation will be conducted for each application.

Completeness is evaluated based on existing omissions and commissions between the nominal ground and the dataset under evaluation.

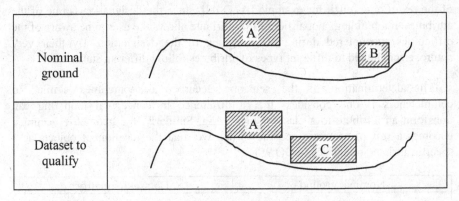

Figure 10.4. *Examples of omission and commission: B = omission, C = commission*

In this context, we can distinguish between two types of completeness (see Figure 10.5), *data completeness*: errors of omission or commission, which are, in principle, measurable and are independent of the application; and *model completeness*: comparison between the abstraction of the world corresponding to the dataset and the one corresponding to the application, preferably evaluated in terms of fitness for use (is the model rich enough to fulfill application requirements? [BRA 95]). Data completeness is broken down into "formal" completeness (concerning the data structure – syntax, adherence to the standards and format used, presence of obligatory metadata), and object completeness, followed by that of attributes and relationships (subordinate to that of the objects). Finally, combining the data completeness with model completeness allows one to estimate the completeness in terms of fitness for use.

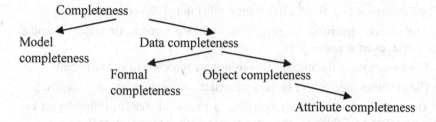

Figure 10.5. *Different types of completeness [BRA 95]*

In summary, completeness monitors the lacuna (omission), as well as the excess (commission) in information contained in the geographic database mainly by answering the following questions [AZO 96]:

– Is the coverage of the zone complete?

– Is the number of objects modeled equal to the number of objects defined in the model?

– Do the modeled objects have the correct number of attributes and are all attribute values present?

– Are all entities represented in the nominal ground represented in the model?

– Is all that is included in the conceptual model also present in the database?

10.3.5. *Logical consistency*

Logical consistency relates to all logical rules that govern the structures and attributes of geographic data and describes the compatibility between dataset items.

Incidentally, this notion was used earlier in data integrity checks for non-spatial data. Its extension to geographic data was done at the time of the first analyses in the domain of topology.

Thus, a dataset is called consistent at the logical level if it respects the structural characteristics of the selected data model and if it is compatible with the attribute constraints defined for the data. There exist several different levels of logical consistency, going from a simple range of attribute values to specific rules of consistency based on the geometry (example: is the contour of a polygon properly closed? [UBE 97]) (see Figure 10.6) or on spatial relationships (constraints of topological integrity – example: every arc of a network should be connected by a node to another arc).

The consistency thus enables us amongst other things to verify that:

– The objects described in geographic database respect the reality (nominal ground) in an exact measure.

– The topology and the spatial relationships are represented and respected.

– The variables used adhere to the appropriate values (limit values, type, etc.).

– The data file is consistent (according to European standards, this aspect can even extend to the reliability of the medium on which the file is stored).

Nominal terrain	Dataset to qualify
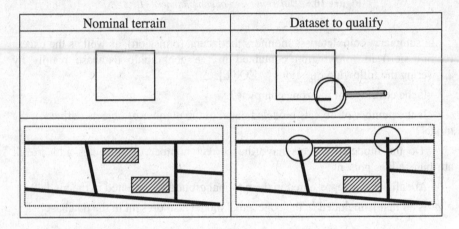	

Figure 10.6. *Example of logical consistency: verification of the closure of the polygon contour, verification of the topological joining relationships*

10.3.6. *Semantic consistency*

The concept of semantic consistency expresses the quality with which geographical objects are described with respect to the model being used. This aspect of quality relates more to the relevance of geographical objects' significance than to their representation [SAL 95]. The semantic relevance is therefore of major importance in determining the fitness for use.

The goal of semantic consistency is to measure the "semantic distance" between geographical objects and the nominal ground. We can, once again, distinguish between the points of view of the producer and the user: for the former, the aim is, on the one hand, to provide documentation on the *semantic content* of his database (mainly by providing the specifications that define the nominal ground, the model, the selection criteria, etc.) and, on the other hand, to provide information on the semantic performance of this database (level of conformance with the above-mentioned semantic constraints); for the latter, the goal is to define the suitability of this data for the user's own requirements. The knowledge of the specifications is, for the user of primary importance, especially from the semantic point of view: do the user and producer agree on a named phenomenon? (For example, does the "hospital" class include clinics?)

As far as the specifications are concerned, two basic levels can be defined [PUR 00]: the geometric level which provides the shape and location of objects and the semantic level to describe the objects. Irrespective of whether the data's physical representation uses a vector model or a raster model, it always respects these two levels: for raster data, the geometry is made up of a collection of pixels and the semantics which are associated with these values; for vector data, the geometry indicates the shape and the absolute or relative position (encoded according to the geometric primitives used) and the semantics bear upon the attributes, their values, or even the explicit relationships between the entities.

The selection criteria define, for example, the input limits (minimum size that a entity should have to be input), operated aggregations, and corresponding criteria ("all crop fields will be stored as "agricultural zones" and merged as required"). The extraction is, finally, a transformation of entities of the real world into objects, attributes, fields of the selected model, and data. To indicate all the parameters used, especially in the generalization procedures implemented, is as important in evaluating semantic consistency as it is for lineage.

In order to evaluate the semantic consistency of a database, [SAL 95] starts by introducing the concept of "ability of abstraction" of phenomena. Some of them are, in fact, difficult to model (boundary of a forest, for example) and it is often worthwhile to evaluate whether the apprehension of the phenomenon is universal or whether it depends strongly on the observer, the context, or the observation date (seasons, shadows, etc.). (See Chapter 5 for a discussion of this problem.)

The methods used for evaluating the semantic consistency can be compared to those for measuring the completeness (omission/commission) of the objects, attributes, and relationships. The semantic consistency also covers the field of logical consistency (data constraints), temporal accuracy (inconsistent dates, etc.), and semantic accuracy (a semantic inconsistency can also denote a classification error, for example) [PUR 00].

In conclusion, semantic consistency is composed of several parameters that cannot be easily differentiated. A flagrant error (for example, a house in a lake; see Figure 10.7) is a semantic inconsistency, but may be due to a temporal error (modification of the banks), a logical inconsistency (not taking into account a house on stilts), or a completeness error (forgetting an island or addition of the house or of the lake).

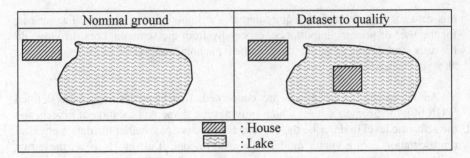

Figure 10.7. *Example of semantic inconsistency*

10.3.7. *Timeliness*

This criterion represents the offset between a produced dataset and the nominal ground on a reference date T. Timeliness provides information about the "freshness" of data. It can be represented, for example, by a validity period for the data batch, a period defined by starting and ending dates.

10.3.8. *Temporal consistency*

The date of the data input, or the date of its revision, is an important factor for the user when judging the data quality (in the sense of fitness for use). Temporal consistency concerns the dates of data acquisition, types of updates, and validity periods.

Depending on the type of phenomenon observed, the management of time-related issues will be different. Some entity classes are re-input at more or less regular intervals (aerial photography campaigns, for example), others require historical management (cadastral maps, etc.); and finally, some are placed between the two types, such as fixed phenomena whose attributes change over time (temperature sensors) or whose location, as well as attributes, can change over time (political frontiers, coastal boundaries). In some cases, the temporal aspect has, therefore, to be treated as an attribute separate from the objects and sometimes modeled as a date, an interval, or a temporal range (validity period) [GUP 95].

We can distinguish three types of time concepts:

– "Logical" or factual time indicates the dates on which the phenomenon, as stored in the database, took place (in reality).

– Time (date) of observation of the phenomenon.

– Transactional time, corresponding to the date the data was entered into the database.

From the user's viewpoint, it is the concept of logical time that is the most important, but in practice, it is the transactional time that is most often stored.

The phenomenon's temporal aspect is highly variable [PUR 00], depending on the type of phenomenon (a mountain's altitude with respect to water level in a reservoir) and the precision with which it was measured. The correct interval for confirming the validity of a database is therefore directly linked to the phenomena that are represented therein. Similarly, the temporal consistency required between objects varies depending on the type of phenomenon: complex entities or ones with inter-relationships require good temporal consistency (topological structures, such as, for example, the road network) whereas independent elements do not require it (sign posts, etc.).

Manipulating temporal information comes down to adding the temporal dimension to the data model used and, by extension, to all the elements of the database, for example, using one or more additional "attributes" for each entity of the database, each attribute, and each relationship. In addition, to maintain a database's temporal consistency, specific mechanisms should be established to allow version-management of data. A modification, such as the segmentation of a stretch of road into two parts, cannot be limited to the removal of the old section and its replacement by the new ones, but should allow the modification of the characteristic of validity of the old object ("anterior", for example) and include the information that the new segments replace the old (to maintain consistency in the history). It becomes obvious that the management of time-related information requires the retention of a large amount of information and dates (modification dates, observation dates, effective dates of updates to the database), and we observe that the management of the temporal aspect can soon become complex, difficult to manage and maintain, and, above all, require large amounts of storage space. The establishment of such mechanisms should be limited as far as possible to those geomatic applications for which it is indispensable.

There exist a number of interactions between the temporal aspect and other quality elements:

– *Lineage*, which provides a lot of temporal information (sequences and processing dates).

– *Geometric accuracy* (for which temporal information can sometimes explain errors).

– *Semantic accuracy* (availability of information on the temporal validity of an attribute allows the detection of inconsistencies when suspect values change).

– *Completeness* (which should only be estimated for entities that are temporally consistent).

– *Logical consistency* (for the same reason).

– *Semantic consistency* (measuring the semantic consistency of the temporal aspects of a database allows the evaluation of the responsiveness of updates to the database with respect to changes in real phenomena).

10.3.9. *Quality criteria: difficulties and limitations*

The quality parameters, or criteria that have been defined, partially overlap each other, which sometimes renders the classification of an error difficult (that is, the determination of which criterion was violated). The example in Figure 10.8, taken from [VAU 97], illustrates this problem:

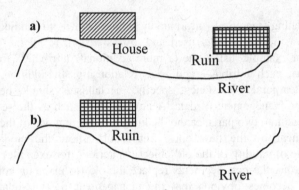

Figure 10.8. *Classification of error cases*

In Figure 10.8, the two datasets represent the same geographical area. The second (b) has one less item. This difference can result from one of three different types of errors:

– An *error* of *geometric accuracy* (the ruin is too far to the left) added to a completeness error (the house is missing).

– A *classification error*, therefore of *semantic accuracy* (the house was classified as a ruin) added to an error of completeness (the ruin is missing).

– A *double error* of *temporal accuracy*. The ruin has disappeared and the house has degraded into a ruin.

The evaluation of quality parameters is useful to the user, but it needs to be easily achieved. In fact, information concerning quality should be relevant so that the users (producers and end-users) accept the constraints that quality evaluation entails and understand its utility. Of course, each producer consciously wants to supply data that is as correct as possible, and each user wants to acquire and use the best available information. Standardization serves as a basis for structuring and evaluating quality, but this basis is still today more oriented toward the data producer than to the data consumer.

The complexity of the standards and, above all, the difficulty in differentiating these quality elements, means that it is expensive to evaluate, store or provide the data quality in a simple and comprehensible manner. Only the evaluation of the gains arising from the use of quality information and a usage that is adapted to the users' requirements can bring home its advantages.

The use of quality criteria mentioned here is a variable depending on the organizations producing and using the data. To facilitate exchange and comprehension of information on quality, standards-developing organizations have published standards which provide guidelines for using quality criteria and for the documentation of procedures for evaluating quality.

10.4. Quality and metadata as seen by standards

10.4.1. *Introduction to standardization*

The goal of standardization, in the meaning of decree number 84-74 of 26 January 1984 and relating to French standardization, is to "*supply reference documents ... solutions to problems ... which arise repeatedly in interactions between partners ...*" Standardization is, above all, an activity of defining specifications in a consensual framework.

Standards emerge from a set of mandated or recognized official organizations.

The French association for standardization (AFNOR) is the motive force behind French standardization and acts as a clearing house for official French, European, and international standardization organizations, whether they are comprehensive in their scope or limited to a given sector (telecommunications, electrical engineering, and electricity), such as:

– International Telecommunications Union (ITU);

– European Telecommunications Standards Institute (ETSI);

– International Standards Organization (ISO);

– European Committee for Standardization (CEN);

– International Electrotechnical Commission (IEC);

– *L'union technique de l'électricité et de la communication* (French National technical union for electricity and communication (UTE));

– European Committee for Electrotechnical Standardization (CENELEC).

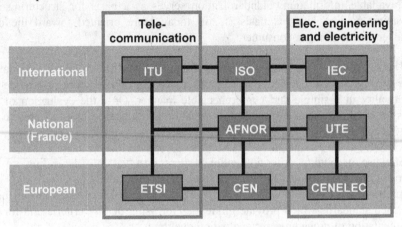

Figure 10.9. *Official standardization organizations*

Around these official standardization organizations gravitate other organizations, often sectoral, self-mandated (but not necessarily less respected), that produce standards in the same consensual framework. Standardization is generally an activity that is the responsibility of organizations that have official status. The expression "*de jure* standard" is often used in this context.

However, standards do not always result from standardization activity. Some specifications take a consensual character without having been designed with such a goal in mind. These specifications are called "*de facto* standards".

These nuances around how standards are finally formed matter little. The importance of standards lies in that they provide answers to problems that arise repeatedly in inter-partner interactions. Thus, as far as quality and metadata is concerned, one has to go beyond individual practices and rely on technical specifications having wide application in the geographical information sector.

10.4.2. *Background of geographic information standards*

The need to exchange geographic information was the motive force behind standardization in the domain. The first standards for exchange emanated from the defense and hydrography sectors in the 1980s:

– The military standard for exchange of geographic data Digital Geographic Exchange Standard (DIGEST) [DGI 00] was established by the Digital Geographic

Information Working Group (DGIWG) which managed and improved it until the early 2000s.

– The exchange standard S-57 [IHO 00] was established by the International Hydrographic Organization (IHO) and is still used for exchanging nautical information destined for onboard navigation terminals.

A little later on, national exchange standards appeared, amongst which are:

– The American Spatial Data Transfer Standard (SDTS) [USG 97], which is one of the precursor standards in the domain.

– The French exchange standard *Edigéo* [AFN 99], which was approved in 1999, after five years of testing.

These different exchange standards implemented to a lesser or greater degree the various quality components. However, their principal defect does not lie in their lack of comprehensiveness regarding these quality components, but in terms of their specific implementations. Each of these standards proposes its own exchange structure within which quality information occupies a specific, but also peripheral, place.

The Content Standard for Digital Geospatial Metadata [FED 98] of the American Federal Geographic Data Committee (FGDC) is a standard dedicated to metadata without being data-exchange-centric. Its goal, defined by the presidential decree 12906 of 11 April 1994, is to capitalize and make available knowledge relating to geographic data produced by American agencies. The importance of quality information is as peripheral as in the data-exchange standards; but the standard's regulatory nature and political will in the US have led to this standard's widespread acceptance and use to this day.

A new approach to the standardization in the domain of geographic information appeared in the middle of the 1990s: one must standardize the different aspects of geographic information and then assemble these standards to respond to different needs (exchange, cataloguing, etc.). It is this new approach that the technical committee 287 of CEN (CEN/TC 287) has chosen by constructing a modular set of standards in the geographic information domain, including, most notably, an experimental standard relating to quality [CEN 99] and an experimental standard on metadata [CEN 98]. The work of CEN/TC 287 came to a premature end with the constitution of the technical committee 211 of the ISO (ISO/TC 211) in 1994.

ISO/TC 211 continued in the same vein as CEN/TC 287, but went much further. After ten years of existence, ISO/TC 211 lists more than 40 published documents of which 75% are standards or draft standards, 15% are technical specifications or draft technical specifications, and about 10% are technical reports. The ISO/TC 211 standards incorporate the application of new information technologies in the domain of geographic information. They create a necessary break between the relational and object-oriented eras, offering new approaches to the entire geographic information community. These standards are modular and, above all, extensible to respond to specific requirements of users while ensuring a sharing of standardized concepts.

In parallel to the work of ISO/TC 211, the Open Geospatial Consortium (OGC) has also established a set of standards in the geographic information domain by taking advantage of the new information technologies. The abstract standards of OGC have strongly influenced the standards of ISO/TC 211, but the originality of OGC arises from its implementation standards such as the format for vector-data exchange Geographic Markup Language (GML) [OGC 03], the specifications for services Web Map Server interface (WMS) [OGC 04a], and Web Feature Service (WFS) [OGC 02], as well as in the specifications for catalogue services, Catalogue Services (CAT) [OGC 04b].

These implementation standards implement abstract standards of ISO/TC 211, notably those relating to the quality and metadata. In addition, these standards are generally taken up by ISO/TC 211 to be published as standards or technical specifications when they are mature enough:

– WMS is the subject of the standard ISO 19128 [ISO 04a].

– GML is the subject of the draft standard ISO 19136 [ISO 04b].

– WFS is the subject of the draft standard ISO 19142 [ISO 05a].

This trend is confirmed by a strengthened cooperation between OGC and ISO/TC 211.

Under the impetus of the project of the European directive INSPIRE, the CEN/TC 287 was reactivated in 2003 to adopt or adapt the standards of ISO/TC 211, thus affirming the importance of these international standards for the European Union. An association of European cartographic agencies and local authorities (EuroGraphics) surveyed its members on their use of ISO/TC 211 standards relating to quality. The survey's results [EGC 04] showed the clear interest that these national agencies have in these standards, but stuttering implementations demonstrated the need for a guide for implementing these standards.

The evolution of standardization of geographic information tends to encourage the joint use of ISO/TC 211 and OGC standards. This general trend does not exclude other standards from consideration, especially those standards relating to the quality and metadata, or even the adoption of alternative technical solutions, most notably:

– The applicability of a solution that is not dedicated to geographic information should be considered before using specific solutions, even if they are of a standard character.

– The OGC and ISO/TC 211 standards should satisfactorily take into account the quality components and metadata both from the theoretical and practical viewpoints.

10.4.3. *Standards relating to metadata and quality*

Quality occupies a prominent and real place in the standards of ISO/TC 211, since three standards and one draft technical specification relate to it:

– The ISO 19113 standard [ISO 02] defines the principles of quality and, notably, of quality components.

– The ISO 19114 standard [ISO 03a] is dedicated to procedures for evaluating quality. It defines the ways of expressing quality measurements, either as evaluation reports or as metadata.

– The ISO 19115 standard [ISO 03b] specifies the conceptual structure of metadata. This conceptual structure takes into account the different quality components defined by the ISO 19113 standard.

– The ISO 19138 preliminary draft technical specification [ISO 04c] describes a set of quality measures.

ISO/TC 211 is still active and other standardization documents relating to quality could still emerge, especially for imaging requirements. The ISO 19115-2 draft standard [ISO 04d] relating to imagery metadata and the ISO 19130 draft standard [ISO 05b] relating to sensor models extend the ISO 19115 standard. In addition, the implementation of these standards requires that other ISO/TC 211 standards, mentioned in section 10.4.5, also be considered.

OGC standards are called upon when implementing ISO/TC 211 standards and, more generally, when implementing services destined for clients more, or less, specialized in geographic information.

Geographic information is, after all, primarily information. It is therefore important to consider general standards relating to metadata and quality. The reference standard for generalized research applications is the Dublin Core [DCO 05] which specifies a fundamental set of 15 metadata items, such as the title, the summary, the date, etc., useful for describing different types of data.

This listing of standards relating to quality and metadata will not be complete if mention is not made of standards in the ISO 9000 series [ISO 00]. They relate to the management of quality and are fully applicable to the production of geographic data.

They allow the incorporation of the evaluation of the quality of geographic data in the more general context of quality control and assurance.

From a strategic viewpoint, the four ISO/TC 211 standards relating to the quality and metadata are therefore used as a complement to the implementation standards in the domain of geographic information, as well as to the general standards, such as Dublin Core and the ISO 9000 series.

10.4.4. *Theoretical analysis of ISO/TC 211 standards*

10.4.4.1. *The ISO 19113 standard*

The ISO 19113 standard focuses on the description of quality parameters. It also calls upon other quality components, such as:

– The use of data in terms of intention (purpose of the data), as well as feedback on the use of data.

– The lineage.

The ISO 19113 standard is mainly descriptive. It delegates the definition of the conceptual structure of quality information to the ISO 19115 standard.

The ISO 19113 standard takes into account the main quality parameters (completeness, logical consistency, semantic accuracy, and positional accuracy) and offers as a supplement a parameter of temporal accuracy. However, the ISO 19113 standard does not address the concept of specific quality, but authorizes the creation of quality elements outside the standardization framework. Such elements can be considered as representing specific quality.

The ISO 19113 standard also proposes a subclassification of the usual quality parameters:

– The completeness is broken down into omission and commission.

– The logical consistency is broken down into conceptual consistency, consistency of the domain of values, consistency of format, and topological consistency.

– The positional accuracy is broken down into absolute (or external) accuracy, relative (or internal) accuracy, and positional accuracy of gridded data.[2]

---

2 The relevance of this criterion is debatable since, on one hand, the type of data representation, ideally, does not impact the classification of quality components and, on the other hand, the differentiation between relative and absolute accuracy is as necessary for gridded data as for vector data.

– The temporal accuracy is broken down into accuracy of time measurement, temporal consistency, and temporal validity.

– The semantic accuracy is broken down into classification accuracy, accuracy of non-quantitative attributes, and accuracy of quantitative attributes.

This subclassification is of interest because the boundaries between the different parameters are typically difficult to define:

– By how much is the measurement of temporal consistency linked to temporal consistency rather than to logical consistency?

– By how much is the consistency in the domain of values linked to logical consistency rather than to semantic accuracy?

These questions illustrate the risks of inconsistency between different implementations of the ISO/TC 211 standards.

Finally, the ISO 19113 standard broaches the subject of some aspects of "meta-quality" without mentioning it outright and without defining the concept.

10.4.4.2. *The ISO 19114 standard*

The ISO 19114 standard specifies a methodology for evaluating quality, the result of which can either be quantitative or be limited to an indication of data conformity to a given specification, which can either be a product specification or a specification expressing the quality requirements of the data for a specific use.

The ISO 19114 standard also defines two methods for evaluating quality:

– A direct method of comparing data with other data, either within the dataset (in this case the method is direct and internal) or external data.

– An indirect method of deducing or estimating a measure of data quality from metadata and, more specifically, from lineage information or data usage.

Whichever is the method used, the evaluation can bear on all or part of the dataset, and can be conducted in a systematic manner on the entirety of the selection or by sampling on a representative subset of the selection.

Finally, ISO 19114 specifies that the evaluation result can be expressed in the form of metadata and/or quality evaluation reports. The standard authorizes an aggregated expression of evaluation results within the metadata; summary results are used rather than detailed results. In such a case, an evaluation report is requested. Quality evaluation reports are covered briefly within Appendix I of the ISO 19114 standard, but no conceptual structure is offered. The result is that this aspect of the standard is often overlooked.

10.4.4.3. *The ISO 19115 standard*

It is paradoxically the ISO 19115 standard which formalizes, in Unified Modeling Language (UML), the quality concepts defined in the ISO 19113 standard and the expression of results of quality evaluations conforming to the methodology defined in the ISO 19114 standard. The experts in the field of quality must feel that their ideas have been appropriated by the metadata experts!

The ISO 19115 standard is organized in metadata sections. The quality information is mainly found in one dedicated section. Some information on the use and timeliness of the data appears in the identification section. Information relating to updating of data is to be found in the section devoted to maintenance. This structure upsets the quality experts, but it ensures a certain consistency of metadata and its use by the users.

A collection of metadata can include several sets of quality data with each set applicable to a selection of the dataset or, more generally, of the metadata *resource* object. Each set of quality data can consist of a set of lineage information and a set of quality-evaluation reports.

The lineage of the resource can consist of information on the sources used and, if applicable, supplementary information on the procedures applied to these sources, but can also very well be limited to a simple textual description. Information relating to the source can be relatively detailed. The resource zone covered by a source can be indicated clearly. On the other hand, it is inconvenient that one cannot specify the resolution of a source image, but can express the scale of a cartographic source.

Each quality evaluation report, called a "quality element", is the expression of the results of the evaluation of a quality indicator. Some aspects of the conceptual definition of these reports are to be noted:

– The quality elements are subject to a classification that follows the classification of quality parameters proposed by the ISO 19113 standard, thus forcing the indicator to relate to a quality parameter.

– The requirement of taking into account specific quality parameters forces the extension of the proposed classification, which, in practice, is somewhat impractical.

– The evaluation result can be expressed in a quantitative and/or a qualitative manner by a simple indication of conformity with a product specification or user specification.

– It is not possible to describe the sampling used for evaluation without extending the ISO 19115 standard.

– The designers of the standard did not want to exclude any type of quantitative result (covariance matrix, for example), thus rendering the expression of a quantitative result somewhat difficult and its use practically impossible in most general cases.

– It is not possible to express an evaluation result in the form of homogenous quality zones. This limitation forces the users to make a dangerous mixture of the ability to select evaluated geographic data and the need to express these zones over which the evaluation result is constant.

However, these problems do not diminish our interest in ISO 19115. Moreover, the establishment of an amendment process for ISO 19113 and ISO 19115 is being discussed within ISO/TC 211 to resolve them.

10.4.4.4. *ISO 19138 preliminary draft technical specification*

On the one hand, the future technical specification ISO 19138 defines information necessary to describe a quality indicator and, on the other hand, provides a description of a list of quality measures. These standard quality measures are, beyond question, a factor for interoperability, but the users' requirements are such that each community should be allowed to describe its own indicators. By standardizing the manner of describing quality indicators, the ISO 19138 standard will allow the emergence of community indicator registries, thus simplifying the approach to quality by organizations for whom the production of geographic data is a secondary activity and who do not necessarily have the means to manage these still-esoteric matters. However, the implementation of such registries will have to wait until this standard, currently under development, has attained a sufficient maturity.

10.4.5. *Standardized implementation of metadata and quality*

10.4.5.1. *Preamble*

Issues of standardization are relevant only during interactions between different actors. By relying on the ISO quality and metadata standards, actors in the domain of geographic information can share credible concepts and principles. To share knowledge of data quality, one has to go further and actually implement these standards.

ISO/TC 211 includes infrastructure standards, such as the ISO 19019 standard which bears directly on the implementation of ISO/TC 211 standards on two axes:

– The model for geographic data exchange impacts the relationship between metadata and the resources concerned by this metadata.

– Only one semantic model (General Feature Model – GFM) governs the manner in which metadata and quality are applied at the level of defining geographic objects.

10.4.5.2. *The model for exchange by transmission*

In the *model for exchange by transmission*, the user invokes services that respond on a case-by-case basis to his queries formulated through a client application. The CAT standard of OGC is the standard reference for services for querying metadata warehouses. It defines the interface between the client applications and the cataloguing service that delivers the metadata. A cataloguing service can be the client of another cataloguing service via the CAT standard.

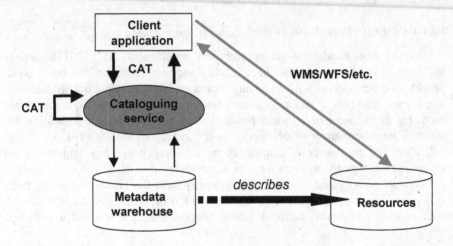

Figure 10.10. *The context for metadata exchange by transmission*

Client applications access resources that are described by the metadata using the OGC access services (WFS, WMS, etc.) or any other non-standard solution.

By default, the CAT standard is based on a subset profile of the Dublin Core which is designed for wide-ranging general applications; but CAT can meet the expectations of specialist applications by offering the possibility of specifying the level of metadata detail expected by the client in the formulation of his query. Within such a framework, it is possible to expect ISO 19115 metadata coded in XML and conforming to the ISO 19139 standard.

10.4.5.3. *Data transfer*

In the traditional *model of exchange by transfer*, the data supplier creates a batch of data that is transferred to the user with the information necessary for its use, most notably its metadata. Several batches of data can be assembled into a set of batches of data having their own metadata. A batch of data contains geographic objects with their own metadata (see section 10.3.5.4).

The concepts involved in the model of exchange by transfer are introduced in the ISO 19115 standard and are detailed in the ISO 19139 standard which proposes XML encoding of metadata and associated resources.

10.4.5.4. *Metadata of geographic objects*

In the sense of the ISO 19109 standard, the relationship between geographic objects and their metadata is ensured through particular attributes of geographic object classes:

– The attributes of type metadata hold complete sets of metadata relating to the concerned geographic objects. Such attributes are accessed by ISO 19115 metadata-consulting applications, the metadata resource being the geographic object with the attribute.

– The attributes of type quality hold the quality elements in conformance with the ISO 19115 standard.

The geographic markup language (GML) standard, the subject of the ISO 19136 draft standard, is recommended for encoding geographic objects and their metadata, for which it is based on the ISO 19139 standard. Moreover, GML is the format used by the WFS and WMS services for accessing data. It can also be used as a format for geographic data in a context of exchange by transfer, conforming to the recommendations of the ISO 19139 draft standard.

10.5. Conclusion

This chapter has presented the growing importance of the concept of quality for the exchange and distribution of geographic data. It has presented the quality components followed by the standards relating to this topic. The most widely-held and used point of view is that of the data producer. Quality and the growing importance of metadata are normally acknowledged and adopted by the producers, whether they are institutional or casual producers. For the users, access to this knowledge is essential. However, we must admit that the "fitness for use" concept, even though not recent, is not really widely implemented. The community must, in the end, develop major types of targeted applications to be able to have the necessary context for evaluating this concept. Tools, indeed even evaluation standards, are yet to be defined, but this new challenge is an unavoidable stage in the evolution of geographic information systems.

10.6. References

[AAL 99] AALDERS H.J.G.L., "Geo-quality and its Impact on GIS Applications", *Proceedings of 21ˢᵗ Urban Data Management Symposium*, 21–23 April 1999, Venice, Italy, p VI.2.1–VI.2.11.

[AFN 99] ASSOCIATION FRANÇAISE DE NORMALISATION, Traitement de l'information – Echanges de données informatisées dans le domaine de l'Information Géographique, 1999, AFNOR, NF-Z52000, p 234.

[AZO 96] AZOUZI M. and MERMINOD B., "Qualité des données spatiales", *Revue Mensuration, Photogrammétrie, Génie Rural*, 1996, Switzerland.

[AZO 00] AZOUZI M., Suivi de la qualité des données spatiales au cours de leur acquisition et de leurs traitements, 2000, PhD Thesis, EPFL, p 107.

[BRA 95] BRASSEL K., BUCHER F., STEPHAN E. and VCKOVSKI A., "Completeness", in *Elements of Spatial Data Quality*, Guptill S.C. and Morrison J.L. (eds), 1995, Oxford: Elsevier, p 81–108.

[CLA 95] CLARKE D.G. and CLARK D.M., "Lineage", in *Elements of spatial data quality*, S.C. Guptill S.C. and Morrison J.L. (eds), 1995, Oxford, Elsevier, p 13–30.

[CEN 98] COMITE EUROPEEN DE NORMALISATION – COMITE TECHNIQUE 287, Information Géographique – Description des données – Métadonnées, March 1998, XP ENV 12657:1998, NF Z52-007, p 62.

[CEN 99] COMITE EUROPEEN DE NORMALISATION – COMITE TECHNIQUE 287. Information Géographique – Description des données – Qualité, August 1999, PrENV 12656:1999, NF Z52-006, p 60.

[DAV 97] DAVID B. and FASQUEL P., "Qualité d'une base de données géographique: concepts et terminologie", *Bulletin d'information de l'IGN* no. 67, 1997, Saint-Mandé, France, p 40.

[DCO 05] DUBLIN CORE METADATA INITIATIVE, DCMI Metadata Terms, http://dublincore.org/ (last consulted: 31 January 2005).

[DGI 00] DIGITAL GEOGRAPHIC INFORMATION WORKING GROUP, Digital Geographic Information Exchange Standard (DIGEST), Parts 1–4, September 2000.

[EGC 04] EUROGEOGRAPHICS – EXPERT GROUP ON QUALITY, Use of the ISO 19100 Quality standards at the NMCAs – Results from a questionnaire, 17 December 2004, Version 1.0, p 15.

[FED 98] FEDERAL GEOGRAPHIC DATA COMMITTEE, Content Standard for Digital Geospatial Metadata, June 1998, FGDC-STD-001-1998, p 90.

[GOO 95] GOODCHILD M.F., "Attribute accuracy", in *Elements of Spatial Data Quality*, Guptill S.C. and Morrison J.L. (eds), 1995, Oxford, Elsevier, p 59–80.

[GUP 95] GUPTILL S.C. and MORRISON J.L. (eds), *Elements of Spatial Data Quality*, 1995, Oxford, Elsevier, p 202.

[IHO 00] INTERNATIONAL HYDROGRAPHIC ORGANIZATION. IHO Transfer Standard for Digital Geographic Data – Edition 3.1 –Special Publication No. 57 (S-57), September 2000, p 114.

[ISO 94] INTERNATIONAL ORGANIZATION FOR STANDARDIZATION, Management de la qualité et assurance de la qualité – Vocabulaire, ISO, Standard ISO 8402, 1994, p 50.

[ISO 00] INTERNATIONAL ORGANIZATION FOR STANDARDIZATION. Systems for Quality Management – Set of standards ISO 9000: ISO 9000, ISO 9001 and ISO 9004, 2004, p 140.

[ISO 02] INTERNATIONAL ORGANIZATION FOR STANDARDIZATION – Technical Committee 211, Geographic Information – Quality principles – ISO 19113, 2002, p 36.

[ISO 03a] INTERNATIONAL ORGANIZATION FOR STANDARDIZATION – Technical Committee 211, Geographic Information – Quality Evaluation Procedures – ISO 19114, 2003, p 71.

[ISO 03b] INTERNATIONAL ORGANIZATION FOR STANDARDIZATION – Technical Committee 211, Geographic Information – Metadata, ISO19115, 2004, p 148.

[ISO 04a] INTERNATIONAL ORGANIZATION FOR STANDARDIZATION – Technical Committee 211 – Working Group 4, Geographic Information – Web Map Server interface – Draft International Standard ISO/DIS 19128, February 2004, p 83.

[ISO 04b] INTERNATIONAL ORGANIZATION FOR STANDARDIZATION – Technical Committee 211 – Working Group 4, Geographic Information – Geography Markup Language – Committee Draft, ISO/CD 19136, February 2004, p 580.

[ISO 04c] INTERNATIONAL ORGANIZATION FOR STANDARDIZATION – Technical Committee 211 – Working Group 7, Geographic Information – Data Quality Measures, Preliminary Draft Technical Specification, ISO/PDTS 19138, November 2004, p 88.

[ISO 04d] INTERNATIONAL ORGANIZATION FOR STANDARDIZATION – Technical Committee 211 – Working Group 6, Geographic Information – Metadata – Part 2: Metadata for imagery and gridded data, Working Draft, ISO/WD 19115-2, September 2004, p 47.

[ISO 05a] INTERNATIONAL ORGANIZATION FOR STANDARDIZATION – Technical Committee 211 – OGC Joint Advisory Group, Geographic Information – Web Feature Service – New Work Item Proposal, ISO/TC 211 N1765, February 2005, p 135.

[ISO 05b] INTERNATIONAL ORGANIZATION FOR STANDARDIZATION – Technical Committee 211 Working Group 6, Geographic Information – Sensor data model for imagery and gridded data, 2nd Committee Draft, ISO/CD2 19130, February 2005, p 160.

[MOE 87] MOELLERING H., A draft Proposed Standard for Digital Cartographic Data, National Committee for Digital Cartographic Standards, American Congress on Surveying and Mapping, 1987, Report number 8, p 176.

[OGC 02] OPEN GEOSPATIAL CONSORTIUM, OpenGIS® Implementation Specification – Web Feature Service – Version 1.0.0, OGC 02-058, May 2002, p 105.

[OGC 03] OPEN GEOSPATIAL CONSORTIUM, OpenGIS® Implementation Specification – Geography Markup Language (GML) – Version 3.0, OGC 02-023r4, January 2003, p 529.

[OGC 04a] OPEN GEOSPATIAL CONSORTIUM, OpenGIS® Implementation Specification – Web Map Service – Version 1.3, OGC 04-024, August 2004, p 85.

[OGC 04b] OPEN GEOSPATIAL CONSORTIUM, OpenGIS® Implementation Specification – Catalogue Services Specification– Version 2.0, OGC 04-021r2, May 2004, p 192.

[PUR 00] PURICELLI A., Réingénierie et contrôle qualité des données en vue d'une migration technologique, December 2000, PhD Thesis, Institut National des Sciences Appliquées de Lyon (INSA).

[SAL 95] SALGÉ F., "Semantic Accuracy", in *Elements of Spatial Data Quality*, Guptill S.C. and Morrison J.L. (eds), 1995, Oxford, Elsevier, p 139–52.

[SER 00] SERVIGNE S., UBEDA T., PURICELLI A. and LAURINI R., "A Methodology for Spatial Consistency Improvement of Geographic Databases", *GeoInformatica*, 2000, vol 4, no 1, p 7–34.

[UBE 97] UBEDA T., Contrôle de la qualité spatiale des bases de données géographiques: cohérence topologique et corrections d'erreur, 1997, PhD Thesis, Institut National des Sciences Appliquées de Lyon (INSA).

[VAU 97] VAUGLIN F., "Statistical representation of relative positional uncertainty for geographical linear features", in *Data Quality in Geographic Information: From Error to Uncertainty*, Goodchild M.F. and Jeansoulin R. (eds), 1998, Paris, Hermès, p 87–96.

[USG 97] UNITED STATES GEOLOGICAL SURVEY – NATIONAL MAPPING DIVISION, Spatial Data Transfer Standard, American National Standards Institute Ed., Draft for review, Part 1–4, 1997, http://mcmcweb.er.usgs.gov/sdts/whatsdts.html (last consulted: 31 January 2006).

Chapter 11

Spatial Data Quality Assessment and Documentation

11.1. Introduction

11.1.1. *Quality in its context*

According to quality management (for example, international standard ISO 9000:2000 [ISO 00]), all the characteristics of a product that arouse requirements about the product are parameters of its quality. Requirements come from actual or possible users, also from other interested parties (personnel, shareholders, production partners, suppliers, neighbors, society at large); they may address the product in its final form as well as the conditions of its production. Frequent quality parameters are: functionality, usability, cost, competitiveness, beauty, availability, size, ethical awareness, user-friendliness, environmental-friendliness, packaging, color, delivering mode, etc. Just as for any other product, geographical data may be assessed for each of these aspects. A twofold peculiarity, however, distinguishes geographical data from other products:

1) Geographical data are more than their material selves: there is more to maps and geographical databases than drawings or collections of bytes. Conversely, geographical data are not reducible to only the information that they carry (which can be said of some other kinds of data, for example, textual definitions in language dictionaries, or markup codes in XML files). While, in other fields of information, data may happen to be whole and complete *transcriptions* of their objects (thus

---

Chapter written by Jean-François HANGOUËT.

acquiring self-referential status), geographical data cannot but evoke their referents. Geographical information (GI) is bound to be a *representation* of the geographic world – a world which, together with its constituents, is too vast, too intense, too diverse, too changeable, proves too complex to be recorded in its entirety. Geographical data are like ambassadors of the world, to all of us who will deal with it on a scale that is too large for direct perception and apprehension.

2) The geographic world (meaning both the extent of geographic knowledge and the continuous challenge to geographical investigation) is exceedingly profuse, intricate, diverse, changing; just like the world in general, it is both magmatic (everything is there, possible or entangled) and phenomenal (objects, however, show themselves and, what is more, repeat their presence). Concerning the fidelity of the representation which is expected from the "ambassador" data, customer requirements are mostly latent (*quality management* speaking), and highly generic ("high tension lines must be reported", "10 m planimetric precision required", "terrain surface must be made continuous with bicubic interpolation", etc.).

11.1.2. *Outline of chapter*

This chapter describes principles and methods used for measuring and reporting the quality of "geographical data" when they are considered under the sole "representation" aspect. Other parameters, which may also prove important for users or producers (such as personnel training, or data volumes), are not addressed here. Representation is isolated from other quality aspects as one of the most original and difficult of all to assess. It lies at the core of exactitude issues when geographical data or applications are assessed. It also comes first when you look for the ontological *raison d'être* of geographic data.

The question "Is the representation operated by geographical data faithful to what should be represented?" is the concern of this chapter.

In order to precisely circumscribe the premier function of any representation – to wit, that of signifying some phenomenon – in order also to equip ourselves with an exact term to comfortably refer to this function throughout this chapter, we first introduce and discuss the notion of *denotation* (section 11.2). The following section reviews possible causes of fluctuations in the denotation, that is, of the gaps that will open between the evident presence (or manifestation) of the phenomenon and its transformation in the hands of data producers (section 11.3). This review will prove useful to coherently justify, circumscribe, and recall the characteristics of various mathematical measures which are usually employed with geographical data to assess the fidelity of their denotation (section 11.4).

11.2. Denotation as a radical quality aspect of geographical data

We use the word "denotation"[1] to designate the original, radical function of any representation: that of being a sign *toward something* (Figure 11.1). What represents is a sign toward what happens to be represented, a sign which, in an essential or caricatured form, expresses both the presence and the intensity of the phenomenon at stake. In our domain, phenomena are *geographic*: recalling the definition given in [HAN 04], they are or can be inscribed ("-graphic") on the surface of the earth ("geo-"). Within this contribution, it is of little importance whether phenomena are concrete or "real" or (since reality "is one among many other possible 'readings' of life", as recalled by [PAV 81]) whether they are immaterial (such as administrative boundaries), fictional (invented lands and countries) or even fantastical (for example, dreams of new, further-reaching borders for one's country). It little matters also whether phenomena belong to the past, present, future, or potentiality. Important is the fact that they may be engraved on the world.

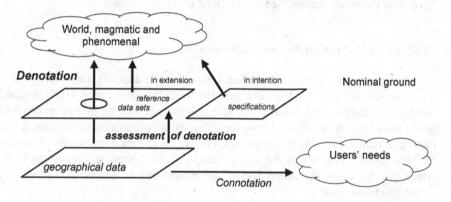

Figure 11.1. *Denotation with geographical data*

The role of denotation is not to point to *secondary aspects* of phenomena, that is, the small details within a phenomenon (which are primary aspects, as constituents of the phenomenon), but everything that comes as a consequence of the existence of the phenomenon (such as its possible usefulness). This new function belongs to what

---

1 The word *denotation* is based on the verb *to note*. The intensive prefix *de-* (not to be confused with the homonymous separative prefix) frees the notion of "notation" from practical contingencies (who notes? how is it noted?) and stresses its essential role by tightening the link with the fact that something happens to be "noted".

linguistics calls *connotation*.[2] When denotation is faithful, connotation is ensured. The secondary aspects of phenomena will not be lost because the users' culture, vision and interest will not fail to summon those aspects in their turn, on the basis of the evidence brought by denotation. The robustness of geographical data to connotation – that is, their aptitude for efficiently serving purposes which differ from the unique concern of representing "the things that are", a robustness also called "fitness for use" in English – proves to be an ever-booming field of research (for example, the works of [GRI 89], [BON 02]). The results of such research could be compared and synthesized only at a meta-methodological level, since the quality measurements involved vary with the kind of applications tested (in most cases with the application itself). It is noteworthy that, in view of the plurality and immensity of the possible uses, in the actual state of research on their systematization, the ISO 19113 standard entrusts connotation not to numerical and quantitative logic, but to textual, qualitative description. The ISO 19113 fields required to document how given geographical data may be used, are indeed of the "free text" kind, in the "overview elements" family (see [ISO 02], section 5.1, section 5.3).

11.3. Sources for the fluctuations in denotation

When geographical data are produced, the resulting denotation may depart from the expected denotation, and consequently from the geographic features which happen to be represented. Sources for fluctuations in denotation may be traced back to the modeling of the world, to the modeling of the production operations, and to the actual realization of these operations. Geographical data being meant to account for some positioning within the spatial, temporal, and thematic dimensions of the world, this is where the differences in denotation can be measured. Let us look closer into these issues.

11.3.1. *The modeling of the world*

The model for the world – that is, what the data should ideally convey – is often called *nominal ground*, in French literature, and *abstract universe*, in English. ISO 19113 uses the expression "universe of discourse" in the same sense: to address all the features from the geographic world that are involved with a given geographical dataset.[3]

---

2 The word comes from *note* and the Latin prefix "*cum*", meaning "with". "Connotation" may be read as: "what comes in the wake of notation".

3 Outside the standards, the expression *universe of discourse* sometimes refers to more than the nominal ground, encompassing not only the geographical features to be represented, but also the producers' and users' habits, viewpoints, cultures, paradigms, technological tools etc.

The nominal ground is usually considered to be instantiated in *intention* (that is, at the higher level of *types* of features to be represented) in the product specifications. Pragmatically, it is often assumed that the nominal ground is instantiated in *extension* (feature by feature) in other available data, considered to be perfect, or whose quality parameters are perfectly documented, and which can be used as reference dataset (Figure 11.1). Note that habits do not reach as far as saying that the nominal ground might be instantiated in extension by the very data to be assessed.

The nominal ground itself is denotative, and designed with imperfections – prejudice, lapses, over-quality, etc. The sole redaction of specifications may result in an erroneous reference, and induce important defects in the data: a misprint in the document, such as *"trees taller than 300 ft are represented"* (instead of *30 ft.*), will deforest landscapes.

Data production consists, first, in modeling the operations to perform, then in applying the operations; both stages are likely to produce fluctuations.

11.3.2. *The modeling of operations*

The model for the production operations (describing what production operations should be applied) might be called, by analogy with the model universe, *nominal production*, *abstract production*, or *production of discourse*. Nominal production is instantiated, in intention, in the production specifications, processing instructions, operation sheets, and other working procedures. Nominal production defines the necessary operations required to produce the right data, and how flows should be organized. A production process is a succession of high-risk stages, from the technology to be employed for data-capture (or for the calibration of source data) down to the methods that will format the output data into the expected computer or (carto-)graphic shape.

Dealing with data transform methods requires awareness of the autonomous dynamics of mathematics when it comes to designing the production of geographical data. It is worth evaluating and documenting the gap between the actual range of what can be achieved and the purity of what is aimed for. Think of the oblate spheroid, which lends itself with more ease to algebraic reasoning than to geometrical computation: the geodetically equidistant, median line between any two points on the spheroid is a crystal-clear concept, yet its explicit equation requires approximation. Such cases recall that mathematics, even on its own ground, may lead to compromise and approximate solutions. In addition, at any stage during production, decisions must be made, indicators measured, statistics computed, thresholds consulted, classifications operated: there again, are the right mathematics employed, the right measurements used, the right regroupings made?

In our heavily computerized context, careful attention must also be paid to computation and approximation algorithms, to the effective calculation processes (which vary with hardware and core-software machinery), and to the methods used to interpret intermediary results (statistics, thresholds).

11.3.3. *Realization of the model of operations*

Planned transforms will be performed: actual results may depart from expectations if input data fail to show the required characteristics, or if instrumental technologies for data capture fall short of nominal performance, if mathematical, algorithmic, statistical operations are not carried out in the right conditions, or with the right parameters.

When transforms and interpretation are entrusted, not to automation, but to a competent operator's expertise, similar fluctuations may occur to data. Advantages of human interaction include the continuous re-centering of numerical drifting, for the operator is able to focus on the peculiarities of every geographic phenomenon to be represented. A possible drawback is that fewer solutions may be attempted in the same period of time.

The eventual inscription of data is also liable to variability, and is another source for the fluctuations of denotation. Misuse, misunderstanding or confusion may arise if data fail to be *legible*, that is, to be clearly identifiable by the target faculties in the conditions of use envisaged – the target faculties being, most of the time, human vision or computer shareability, yet also touch (tactile maps) and hearing (acoustic spatial information [MÜL 01]).

11.3.4. *Realization of the model of the world*

The model of the world comes to instantiation through the realization of production operations. To measure their impact on denotation, what is to be observed?

Following an ontological approach, worldly things show characteristics: provided these characteristics are correctly rendered in data, denotation is faithful. In order to organize the huge variety of ontological possibilities among features, we have recourse here to an ontological approach, yet not to worldly things, but to their global container, *viz* the objective world, which is reputed to have three "dimensions": *space, time* and (according to the terminology used by philosophers and novelists alike, such as Nietzsche and E.M. Forster)[4] *value* or *theme* (more frequent in geomatics).

Things are being materialized in this world: the world may be seen as an accommodation structure for the things. This is a subtle structure which Plato, in his *Timaeus*, calls by the name of *khôra*, "place", and describes as "the receptacle, and as it were the nurse, … the substance which receives all bodies … [and which] is invisible and all-shaped, all-receptive" (*Timaeus*, 49A, 50B, 51B). (*Field* theories in modern physics and mathematics are closely related to this view.) There is a counterpart in the representation: the accommodation structure inherent to the model. This is as subtle and implicit, and indeed as fundamental, as the world *khôra*. For every representing thing to take shape, the receptacle of the representation must show some consistency: space must be space (often mathematized as Euclidean geometry), time must be time (usually linear, expressed in a regular calendar), value must be value (by reproducing the field of values identified in the world).

Habits in geomatics drive us to split each of these three aspects in two, according to the metric/non-metric viewpoint (for example, geometry/topology). The latter, the de-meterized, qualitative, "articulatory" concern, addresses the objects' properties which remain invariant when space (or time, or theme) is continuously deformed (without breaks or folds) and deforms in turn the objects immersed within it.[5]

This, epistemologically speaking, induces another constraint on the possible types of accommodation structures for the model: the kinship with the *khôra* of the world must be even tighter. The whole articulatory background of the model (topological, chronological, or semiological) must be continuously deformable; there must be homeomorphism with the articulatory background of the world.

Let us describe in a few more details the metric/non-metric sides of the three dimensions of the world.

Space: geometry and topology

Geometry is the numerical location, relatable to some absolute coordinate system, of objects, groups of objects, and shapes. Is the represented feature at the right place in the spatial background of the model? And with the right proportions?

---

4 See, e.g., F. Nietzsche's *On the Genealogy of Morality* (1887), and E.M. Forster's *Aspects of the Novel* (1927).
5 This generalizes the definition of mathematical inspiration usually given by geomatics to "topology". See, e.g., [ISO 03b]: "Topology deals with the characteristics of geometric figures that remain invariant if the space is deformed elastically and continually".

Topology[6] is what remains invariant in the geometrical properties of objects when it is envisaged to continuously deform the space they are nested in. In a way, in a very general sense, topology is thus what remains of the geometry of objects, individually or in configuration, when they are no longer measured: "join", "are disjoint", "contains", "is contained by", "follows", "precedes", etc. Topology in geographical information systems (GISs) is often seen as the logic behind *crossings*, *ramifications*, *inclusions* and *continuations*. This is because the geometrical properties observed about the objects – those properties which, according to the definition, must resist the (mentally performed) continuous deformation of the *khôra* – are historically limited to the graphs and regions directly deduced from the joining together of the objects' geometrical traces (road-, railway-, river-axes, etc., field or administrative boundaries, etc.). Today's progress in geomatics is toward a wider and more sensitive kind of topology, where graphs and regions to be elastically deformed result from more elaborate constructs than a mere joining up of traces. Topology thus extends its reach by being able to account for *remoteness* (using Voronoï diagrams of the "furthest" kind, [OKA 92]), for *alignments* and other *visual clusters* (using minimum spanning tree and similar graphs, [ZAH 71], [REG 98], or more generally Voronoï diagrams of the usual kind, [HAN 98]), for *mutual integration* of objects in their set (as can be evaluated with the curious "relative adjacency" of [BER 03]), etc. Non-metric, that is, topological, information may thus be expressed, to qualify both shapes and arrangements: "ahead", "further behind", "next to", "unavoidable", "to the left", "to the right", "in the distance", etc.

Concerning the quality of topological denotation, issues at stake include: are the relative locations of the represented objects qualitatively faithful to the configuration of the objects in the world? Is the inner organization (necks, self-crossings, portions in succession) of a represented object faithful to that of the worldly object?

Time: date and chronology

In the representation, both date (the dating of the objects in the world: moment, duration, rhythm, etc.) and chronology (tempo, simultaneity, concurrency, chaining, temporal inclusion) may be of variable quality.

Is the date given with the data faithful to the (current, historical, projected) date known for the object in the world? The worldly object goes through various stages: those of its lifecycle (for example, for a building: erection, possible extensions, transformations which may include eventual demolition), yet also those of its

---

6 Although ISO 19113 subsumes topology under a highly composite "logical consistency" ([ISO 02], section 5.2.2), topology is explicitly considered here as one of the aspects of geographical denotation.

possible spatial courses, when the object is mobile or movable. Are these stages correctly dated in the representation?

Note: what is called here by the name of *chronology* is to date what *topology* is to geometry. The similarity is so close that many authors in geomatics use the expression "temporal topology" to designate the "articulatory" aspect of time, and work with "temporal graphs" that are homologous to these "planar graphs" which happen to be so handy when it comes to transcribing topology.

Value: semantics and semiology

"Semantics" addresses the meaning conferred on the object or group of objects identified in space and time, that is, the expression of its thematic value, which ranges from its geographical identity to its particular name, via its recognized role or function, or the description of an inherent flavor or color, etc.

"Semiology"[7] is meant to cover juxtapositions, inclusions, continuations, confrontations, thematic redundancies between the values of objects or of pieces of objects within themselves. In object-oriented models, semiology is often instantiated with the actual definition, composition, association or constraint relationships between classes. More generally, semiology may be said to be instantiated in the actual realization of the *formal ontology* which happens to be (explicitly or implicitly) defined for the data.

That building, which happened to be encoded as "house" in the database, is it really a house in the world? The area where the building is contained, is it really thematically part of the more general district where the building is contained? Such questions belong with the semantics and semiology of denotation.

11.3.5. Synthesis

The figure below summarizes the possible causes for the fluctuations of denotation as described in sections 11.3.1 to 11.3.4 above. The faithfulness of a

---

7 By their usual interpretation, "semantics" and "semiology" deal with *linguistic signs*, the former domain addressing the meaning of signs, the latter the dynamics of sign systems. Our definitions generalize their respective ranges, to meet some of the wider branches of Peirce's or Eco's semiotics. In addition to the fact that data are signs – and thus susceptible to semantical/semiological considerations – the worldly objects themselves are given values (because their phenomenal manifestation is noted, or because they draw even more specific attention to themselves once the nominal ground is designed); these objects become extracted from the magmatic world and are made to carry meaning, thus accessing "sign" status.

denotation, for each of the three generic dimensions (space, time and value), results from the quality of the modeling and of the transforming operations involved in the production. The figure also shows that denotation faithfulness may be measured at any stage. Conversely, as genealogical reminder, it may help producers to document each of the effects that methodological and technological decisions may have on data quality.

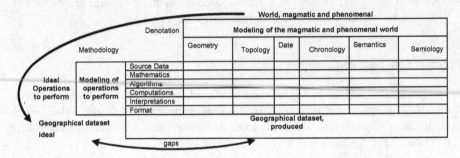

Figure 11.2. *Gaps in denotation*

11.4. How to express denotation quality

Denotation quality may be assessed, yet raw results bear little significance if the method by which they are produced remains obscure. Notably, the power of a measure, which may be deduced from – or induced by – its name, is not to be confused with its effective signification, which is ensured by the measuring method. The first paragraphs in this section review the issues specific to *measure* that influence, if not denotation quality itself, at least its observability.

11.4.1. *General principles for the assessment of denotation quality*

Five general principles regarding the assessment of denotation may be counted: Assessment is based on *comparison* (of produced data to the world); elementary *measures* on objects and *statistics* are made, validity of measured characteristics may be pronounced on their inclusion within *confidence intervals*. Eventually, *reporting* of the results is required.

11.4.1.1. *Comparison*

Assessing denotation quality requires comparing data that are produced with data that should be produced. The latter being of ideal condition, reputedly more precise data are used in their stead for reference: either available data or data especially captured for the occasion through adapted surveying methods.

When the reference data over-cover the spatial, temporal, and thematic ranges of the produced data, comparison may be total; most of the time, however, comparison is but partial. In order to confidently extrapolate the conclusions of partial comparison to the whole dataset, it is first important to choose with care the objects and characteristics to be compared, and the measuring methods to employ: this is the domain of *sampling*, as studied in, for example, [ISA 89], [BON 02]. Appendix E of ISO 19114 [ISO 03] presents different strategies to define samples, relying on judgment, or on statistical procedures which differ as to the kind of control they bring over the general characteristics of the samples:

– *Simple random sampling* is the purely random selection of objects and characteristics to be assessed (which may result in geographically meaningless samples).

– In *stratified random sampling*, data are first divided into geographically meaningful sub-populations (following some well-defined spatial, temporal and thematic criteria), and sampling is performed in each sub-population.

– In *semi-random sampling*, objects are randomly selected not from the entire population, but from those objects which happen to be located at pre-defined positions (in space, time, or themes); for example, the random selection of houses from the houses found around the knots of a regular grid superimposed over the town. With this method, data may be picked up from all over the (spatial, temporal or thematic) full extent of the dataset.

Comparison involves another crucial step: once both the dataset to be assessed and the reference dataset are available, what can be compared must be determined before measurements can be made. This is a matter of *data-matching*, an operation which, either automatically or interactively, aims to fix the (reference object, object to assess) pairs before measuring the mutual differences [DEV 97]. The difficulty of data-matching cannot be overestimated: there is seldom mathematical bijection between the *objects of the world which happen to be represented* in the two datasets and, even if bijection between the objects were the case, bijection *between the objects contained* in the two datasets would be exceptional. Data-matching is no deterministic operation; an object in the representation (say one of a series of hairpin bends on a generalized map) may correspond to any one of several objects in the reference (some bend or other in the actual succession of bends). Data-matching requires a continuous aim for the geographic phenomenon through its representation, which calls for unflagging attention when the operation is performed, or when automated methods are designed [HAN 04]. It may also be added that data-matching is no innocuous operation; automated or interactive, the decisions to be taken will, of course, influence subsequent measurements.

11.4.1.2. *Measure for measure*

Quality measures can be invented easily, effortlessly; for example, compare the number of vertices used to account for a contour, or compare the areas within contours, or compare character strings for names, etc. – once spontaneously suggested, their meaning, however, requires further consideration. Conversely, when it comes to assessing a particular aspect of denotation, what measure can be used?

Look at geographic location: to measure distance between represented and representing objects, should you use Euclidean distance? Hausdorff distance? Fréchet distance? etc., plane geometry? or spherical? or spheroidical? etc.; should you use a L_2-, L_1-, L_∞ norm? or another norm? What geometric pieces of the objects to be compared will you consider? Their centroids? Their contours? Their whole material bodies? Will you transport the object to be represented into the coordinate system used for the object that represents, or operate the reverse projective transform? Also, are such simplistic measures legitimate? Notably for reasons related to cartographic generalization [HAN 98], the geographic location of an object may be largely inaccurate in absolute coordinate systems, while correct and true, relative to its neighbors.

A measure, as a complex whole, is neither understandable nor meaningful when reduced to a number or a formula. The name given to the measure must be evocative (by itself) and consistent (relative to other measures that are or could be used). Finding a correct name for a measure is not that straightforward. Its principles, including the mathematics employed, must be explained, as well as its own precision. Because different algorithms may provide different results, the principle of the algorithm used is worth description, sometimes even the sensitivity of the calculus machine.

11.4.1.3. *Statistics*

Statistics in denotation assessment intervene at three levels:

– Always, as soon as a few objects are involved, in order to present comparison results in a way that is synthetic or tractable enough for human apprehension.

– Quite often, in order to generalize the meaning of measures taken on a few data samples to characterize whole datasets.

– When a measuring method is designed, notably by researchers, in order to evaluate its intrinsic precision (so that it can be used advisedly ever after).

In the first case, the profuse measurement results are usually summed up in a couple of figures, the most frequent *mean* and *standard deviation*. For these values to be significant, comparison results must be in sufficient numbers. One hesitates to put much confidence in the significance of a mean computed on say three or four values. The reflex-threshold in statistics is 30 values, yet this relies on statistical models which may prove questionable for geographic purposes [DAV 97].

The second case, that of statistical inference, raises difficult and critical issues which are still objects of ongoing research (see, for example, [BON 02]). What distribution or regression models should be followed? Which formulae? How should samples be selected? What grounds may account for the adequacy between the *mathematical* dynamics proper to the sample (that is, the mathematical assurance that results mathematically inferred or extrapolated will be mathematically sound) and its *geographical* dynamics (its representativeness, the legitimacy for its serving as a model for the whole representation)? The ISO 19114 standard provides guidelines for sampling strategies [ISO 03], yet it has never been proved that geographic objects systematically satisfy the conditions required ` for the mathematical validity of the statistical operations and approaches one may be led to use [BON 02].

In the third case – the analysis of the reliability of measuring methods – the repetitiveness of measuring methods is tested. When measuring is performed n times on the same object, how do results vary? A computerized measuring method, applied to a same object, will always result in the same value. To estimate the inherent precision of the measure that is being designed, statistics are used to simulate its behavior through small perturbations brought to the objects themselves, and to observe how results do vary. In the case of denotation measures, perturbations may be generated in both the reference dataset and the dataset to be tested, which poses methodological difficulties. What is more, the geographical objects involved, in life-size datasets, are contingent and variable, in the eyes of mathematics, and the extension of the precision of the measure (as defined at the conception stage) to its possibilities in real applications raises difficult issues [ABB 94], [VAU 97].

11.4.1.4. *Validity intervals*

Thresholds and intervals used to assess data validity are usually identified at the conception stage and recorded in the specifications of the product. They provide a first, very general, overview of quality; when the measure for some aspect falls in the right domain, the aspect "passes the test". When several aspects are measured, some will pass, and some will fail. How can some indication on the general quality of data be given? Rules must be adopted. You can require that every test be passed before stamping the data as "good". When tested aspects differ in their importance, the weighted average may be preferred (see ISO 19114, Appendix J on "*pass/fail*" tests, [ISO 03]). This being recalled, three methodological precautions may be insisted upon here.

In order to test the validity of data or datasets, the precision appropriate to the measuring methods must, of course, be smaller than the thresholds and intervals.

When an interval holds for a *set* of data, it is not always possible to proceed by *separately* measuring the differences for the *n* objects in the set. Or, at least, the formula for synthesizing the *n* differences must be mastered, so that the unique figure may be theoretically and effectively comparable to the interval. A road network, for (counter-)example, may be thought to be well positioned as a whole, relative to the network given in the reference, because traces superimpose themselves – and yet some individual sections may well be captured far from their actual places, chance only bringing them close to other sections in the reference.

Thirdly, two kinds of thresholds or intervals are usually distinguished: thresholds setting out users' requirements or producers' commitments (usually called *requirement thresholds*), and thresholds that are likely to be satisfied, known from various studies, including the analysis of what production lines can produce (*target thresholds*). The importance of *pass/fail* tests should be interpreted accordingly.

11.4.1.5. *Reporting*

Assessment of denotation quality, as a (meta)data production process, suffers from the same imperfections and compromises that tend to alter the representations themselves (see Figure 11.2). Mathematics, algorithms, effective computations, methods, naming of operations: between what is expected to be measured, and what is effectively measured, gaps open and the rendering of quality is deformed, augmenting the deformation of denotation.

For denotation quality of geographical data to be fully reported to users, raw data and mere results are insufficient. Additional elements are desirable, so that the meaning and validity of the measures themselves can be appreciated, such as the essential characteristics of the measures employed (for example, their inherent precision), the proportion of data tested, the proportion of data that could not be tested (for whatever reason, for example, because no counterpart was found in the reference), and more generally the metadata that are useful to portray context and purpose (kinds of objects tested, date of testing, person in charge of measuring, reference of product specifications, hardware type, etc.). To convey the results of the measures, mathematical synthesis is but one of several possibilities. Notably, when myriad observations cannot be distilled into some ultimate figure, schemas and visual portrayal are preferred. Their own clarity and fidelity to the measuring methods and results rely for the larger part on the command of Bertin's "semiology of graphics" [BER 67]. Reports sometimes take the form of "quality maps" where "ordinate visual variables" are used to show the degree of imprecision attached to each object. Complementary clarification, expected in addition to raw results, may consist of discursive analysis, or of the genealogically-structured compilation of what denotation gaps are made of (a stair-by-stair climbing up the "methodological" column in Figure 11.2).

11.4.2. *Toward a few measures*

These last few sections, using the three-fold ontological classification of the world in space, time, and value, review a few measures which are commonly used in geomatics to assess denotation gaps between produced and reference data. Documenting, in detail, all possible quality measures defies feasibility; the following list is provided as an illustration of possible measures, and of precautions that may be taken to preserve the intrinsic meaning of the measures.

Note: the term "objects" in the following sections designates geographic entities which are either identified upon storing in the database or consequently recognized by the naked eye, or hand, or by algorithms (whatever the type of database: vector, documentary, raster, image). Geometrical computations differ, whether data are in a vector or raster form; the distinction however is not important in this general discussion.

11.4.3. *Geometry measures*

For convenience, we will use here an X-Y-(Z) Euclidean, ortho-normal coordinate system to structure the space of representation. Although a common system, it is not the only possibility.

11.4.3.1. *Punctual objects*

For point-objects – those objects whose geometrical trace is recorded as a point – measuring location differences usually amounts to calculating distances between produced position and reference position (possibly separately along each coordinate axis, in order to elicit potential biases). For synthesis, *mean error* is usually computed: harmonic, geometric, arithmetic or quadratic mean (the last resulting in larger values, is the most "demanding" of all, for the same quality threshold). Quadratic mean computed on the XY plane is usually called *planimetric quadratic mean error* [DAV 97]. The *regular grid for regional bias* is used to show local deformations within the whole data extent. A regular grid is (fictitiously) superimposed on the dataset to be assessed and, at each knot, mean values of the differences in (X, Y or Z, etc.) positions are computed on the (tested and reference) objects that are closer to the knot than to any other knot.

To give such measures their full meaning, complementary indications must be provided, such as *sample size*, and *rejection rate*, that is, the proportion of tested data for which no homologous object could be found in the reference when comparisons were made. The proportion depends on the tolerance threshold allowed for matching around each object; this threshold also must be provided (and justified).

11.4.3.2. *Linear objects*

Linear objects, the trace of which is a line, share with surface objects the particularity of showing highly irregular geometric forms. In vector GISs, line-objects are usually stored as successions of segments.

One of the most frequent distances used to assess the difference of location between two lines (tested line and reference line) is *Hausdorff distance* (Figure 11.3b; for a definition, see [HAN 98], [BAR 01]). Finding the value of the Hausdorff distance between A and B requires finding the largest of the smallest (Euclidean, most frequently in literature) distances *from* A *to* B, and, reversely, the largest of the smallest distances *from* B *to* A. The larger component is the Hausdorff distance.[8]

This distance may be computed on closed lines, and more generally on point sets of any kind. It is a mathematical distance, thus having interesting properties. Most notably, when its value is null, the two sets of points are exactly superimposed (a wrong deduction when, say, the smallest possible distance between two sets is chosen; two lines, crossing, will show a "null" distance, while not similar).

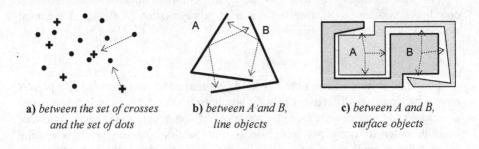

a) *between the set of crosses*   **b)** *between A and B,*   **c)** *between A and B,*
    and the set of dots   *line objects*   *surface objects*

Figure 11.3. *The two components of the Hausdorff distance*

---

8 Lines (or surfaces) involving an infinite number of mathematical points, an infinite number of distances, should be computed to find Hausdorff, according to this definition. Practically, this may lead to using approximate methods. On vector data in habitual GISs however, Voronoi diagrams make exact computation possible, by reducing the problem to finite proportions (see [HAN 98] for a discussion and fundamental references). On raster data, classical image processing techniques, such as "dilation" (see [SCH 93]) make Hausdorff computation easy, the *A* to *B* Hausdorff component providing the solution to the question: *"what is the minimum dilation of B that entirely covers A?"*.

Results read easily:

– HD_1 being the Hausdorff value between A={Data} and B={Reference}, means that data are globally no further than HD_1 from the ideal positions.

– If any Hausdorff distance between A={One Object$_i$} and B={The Reference For Object$_i$} is less than HD_2, then the Hausdorff distance between A={All Objects} and B={All References With A Representation} is at most HD_2.

– The reciprocal assertion is not always true; an object may be positioned very far from its own reference, but very close to another reference.

From the differences in the positions of lines, as measured Hausdorff-wise, [ABB 94] elaborates a kind of "linear planimetric quadratic mean error", which takes its full sense in a well-defined, particular methodology. A statistical model for the precision of linear objects is described in [VAU 97].

Another mathematical distance that befits lines is the *Fréchet distance*. This requires the lines to be oriented. The exact definition is quite laborious, yet can be expressed intuitively with an image. On one of the lines trots a dog, on the other the master. They are allowed to stop, concurrently or not, yet neither is allowed to walk back. The Fréchet distance between the two lines is the minimum length of the leash that makes such combined progression possible [ALT 95], [BEL 01]. An implementation of the Fréchet distance for matching purposes is proposed by [DEV 02]. On closed lines, such as lake or building contours, several computation cycles must be made, with varying starting points [ALT 95].

The Hausdorff distance between lines, which tends to convey land use differences, is always smaller than or equal to the Fréchet distance, which conveys shape differences (Figure 11.4).

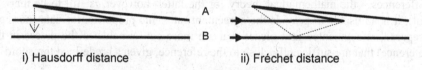

i) Hausdorff distance   ii) Fréchet distance

Figure 11.4. *Hausdorff and Fréchet distances on lines*

Together with the results of these measures, it is advisable to provide the *agreement rate* between the geometric traces in the tested dataset and the traces in the reference. This rate depends on a validity threshold to be provided and, ideally, justified. The agreement rate is the proportion of data for which differences are acceptable according to some predefined threshold. The method requires preparation of the data first, because the objects to be compared may not be segmented identically in the reference and tested datasets; it would not be fair to measure the difference in position of a line captured "with a tail", on the one hand, and that "without a tail", on the other hand (Figure 11.5). Of course, the rate of re-segmented objects has to be provided, and the specific principles and methods used for re-segmentation have to be clearly described and explained.

Figure 11.5. *Hausdorff (or Fréchet) distance on non-prepared lines*

11.4.3.3. *Surface objects*

"Surface objects" are filled with their own matter; the geometrical being of a lake is not only the trace of its shoreline, but also the extent within – even if vector GISs will store little more than a succession of well-chosen vertices on its contour.

Geometric measures used to compare surface objects with their references usually make do with contours only. They are often inherited from image processing techniques for shape recognition. A wide range of methods, applied specifically to the measurement of shape differences in geographical surface-objects, is given in [BEL 01]: "polygonal signature" (that function which, to any point on the contour, associates its distance from the centre of mass of the surface); "areal distance" (the total area of those parts of the two objects that are not in the intersection of their surfaces, divided by the total area of their union. This measure can be computed easily, and proves quite robust, notably for data-matching purposes.); "Hausdorff distance" again (yet between full surface objects and contour-only approaches, results may differ, as illustrated in Figure 11.3c), etc.

Again, together with the overall results (arithmetic or quadratic mean of the differences – the mathematical theory for the latter, however, is still to be mapped out), it is advisable to provide the agreement rates: the proportion of data (possibly prepared, re-segmented into pieces that are more conformable to the pieces in the reference) that are sufficiently close to the reference, given a predefined threshold.

11.4.3.4. *Topology*

As topology in geomatics is, most of the time, expressed in the form of graphs, graph-similarity measures can be used to assess topological differences in geographical data. Graph-similarity assessment is a complex and difficult domain, for which [BOL 02] provides a good introduction. A simple measure can be mentioned here: the Hamming distance, which amounts, in its most basic version, to counting the number of arcs (or nodes, or faces, be it all of them, or those showing special characteristics) that happen to differ in the two topological structures. Results read easily: the smaller the value, the more similar the topology. This method requires preliminary topological matching, the principles of which have to be provided with the results.

For results somewhat more informative than Hamming, *excess* and *deficit* rates (see section 11.4.5) of nodes (or arcs, or faces) can be given.

11.4.4. *Time measures*

11.4.4.1. *Dates*

Measures used for point geometry can also be used for the assessment of dates: one-dimensional means of various kinds, sample size, rate of values outside the nominal temporal interval, etc. In addition, there are measures for distances between intervals, so that duration differences can be quantified. One of these is similar to the Hamming distance, above, and to the areal distance (let us call it "durative distance"): the total time of the durations that do not happen at the same time between object and reference, possibly divided by the total time of common and non-common durations (Figure 11.6). As to rhythm measures, they will bear upon frequency differences.

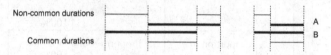

Figure 11.6. *Common and non common durations of two objects*

11.4.4.2. *Chronology*

Just as with topology, differences in chronology may be measured in terms of deficit and excess of temporal nodes or arcs. In more elaborate forms, assessment will have recourse to graph-distance techniques. Notably, [OTT 01] and [ALL 91] present other time-related issues which may be worth measuring in GISs.

11.4.5. *Value measures*

11.4.5.1. *Semantics*

Semantics is often instantiated in GISs through denominations, that is, by names for things, roles, and attributes. Assessment of semantic differences between geographical datasets, however, does not bear (yet) upon the distance between the *meanings* of words (as studied by psycho-linguistics in the wake of [OSG 57], by lexicographic analysis, or more generally by natural language processing). This is only a recent area of research in geomatics (see, for example, [KUH 03], [ROD 03]). Habitual measures are proportions of correct and incorrect denominations, the expertise as to the proximity of meaning being entrusted to the human operator (or, in highly systematic cases, to suitably automated routines).

The *excess rate* is defined as the proportion, *in the data*, of elements of the same aspect (belonging to a particular class, sharing similar values for some textual attribute) which exceed what exists in the reference. The *deficit rate* is defined as the proportion, *in the reference*, of elements of the same aspect (*idem*) that are not represented in the dataset.

The *confusion rate* between aspects i and j ($i{\neq}j$) is the proportion, in the reference, of elements of *aspect i* (in the reference) which happen to be represented as *aspect j* (in the data). The *agreement rate* is defined analogously, imposing $i{=}j$.

These measures are regrouped in the so-called "confusion matrix" (Table 11.1).

Data ⟋ Reference	...	i	...	Aspect j	...	None (reference without homologue)	
...	...	...	...	...	...	...	$\Sigma_j + _{none} = $ 100%
Aspect i	...	...	...	Proportion of items under *Aspect i* in the reference classified as *Aspect j* in the data	...	Proportion of items under *Aspect i* in the reference without homologue in the data	$\Sigma_j + _{none} = $ 100%
...	...	...	...	...	...		$\Sigma_j + _{none} = $ 100%
j	...	...	...	...	...		$\Sigma_j + _{none} = $ 100%
...	...	...	...	...	...		$\Sigma_j + _{none} = $ 100%
None (data without homologue in reference)	...	...	...	Proportion of items classified as *Aspect j* in data without homologue in the reference	...		*excess*
...		*confusion elements*				*deficit*	Diagonal: agreement elements

(To the right of the table, spanning the rightmost column: *confusion* ... *elements*)

Table 11.1. *Confusion matrix*

The confusion matrix is sometimes also called a "misclassification matrix" in the literature, and may be filled with total numbers, instead of proportions. From the measures it gathers, other indicators may be deduced, such as *producer's accuracy* (proportion of reference elements correctly classified, per reference class), *consumer's accuracy* (proportion of data elements correctly classified, per data class), or the Kappa index, which provides a unique (if pessimistic) number to express the overall certainty of the classification (see, for example, [GAM 95]).

Attributes, according to the data model or specifications, may occasionally be left unvalued. The assessment must distinguish, consequently, amongst: non-valued attributes that conform to the reference (because the attribute is *missing* for the particular object found in the reference); non-valued attributes that are in contradiction with the reference (*deficit*); and attributes which are valued while no information can be found in the reference (*excess*). The various rates concerning an attribute are classically gathered in a "presence-absence matrix" (Table 11.2), to which is added the indication of the size of the samples tested. [GAM 95], [DAV 97], and [FAÏ 99] propose other measures.

Reference \ Data	Not valued	Valued
Not valued	*Absence rate*	*Excess rate*
Valued	*Deficit rate*	*Presence rate*

Table 11.2. *Presence-absence matrix for an attribute*

11.4.5.2. *Semiology*

Semiology is often instantiated in geographical datasets through the relationships and associations specified by data models and, more generally, by ontological models. One of today's greatest issues in geomatics lies in ontology matching [MAE 01], [FON 03], which is often approached using as a comparison of graphs. For the time being, the resolution of semiological differences is entrusted to human expertise (either to the operator who is in charge of controlling data, or to the designer who will program routines when comparison is known to be mechanical), and measures accounting for semiological differences will include, as with semantic measures, agreement, deficit, excess and confusion rates between the reference and tested representations.

11.4.6. *Indirect measures*

In addition to direct measures of denotation differences, that is, those measures that compare data to some reference dataset, indirect measures must be mentioned here. Data are no longer compared to the nominal ground in extension, but to the nominal ground in intention.

The nominal ground in intention, for example, in the form of specifications, aims to calibrate the objects in the world to be integrated in the data to be produced. Indirect assessment of denotative quality thus deals with verifying the coherence of a representation within the generic model that predicted it, not with its worldly version. The question is no longer, say, "*Is this tree really 100 feet tall?*", but "*Is the value of the 'tallness' attribute of this 'tree' object comprised between 10 and 60 feet as predicted by the ontological model?*". The ISO 19013 standard, and indeed current habits in relation to quality in geomatics, rank such issues under the notion of "logical consistency". Because they do not require effective comparison with the world, these are autonomous measures, which can be made exhaustively and automated systematically. Their results will take the form of agreement rates, of violation degrees, etc.

11.4.7. *Measures on modeling*

The various measures described above are meant to be made between the dataset to be tested and a more precise dataset taken as reference. As shown in Figure 11.1, this, however, is surveying but one part in the path of denotation: from the representation to assess to another representation, if a more precise one. The magmatic, phenomenal world remains in the distance.

Figure 11.1 encourages reducing the complementary path, by assessing the quality of the specifications themselves (where the nominal ground is brought in intention), also by assessing the quality of the reference data, using yet another nominal ground, in extension or in intention.

For the most part, these new stretches in the domain of geographical denotative quality are still unexplored. How should the quality of the nominal ground given in extension be expressed? And that of the nominal ground given in intention? How do both nominal grounds, in extension and intention, fit with each other? How can yet another, more precise, better adapted, nominal ground be created? Supposing these issues are solved, what operations, what methods, can be used to feed quality measures made on datasets with the knowledge gained on nominal grounds? By the very principles it poses, the assessment of geographical denotation quality opens vast vistas of research.

11.5. Conclusion

In addressing the "representation" aspect of geographical information quality, this chapter has first conducted a structured review of the possible fluctuations in data denotation. From the specifications of the data to be produced, down to the measures themselves used to assess their quality, with every modeling and producing stage, the irreducible gap between data (that represent) and world (that is represented) keeps on deepening and widening. Measures liable to account for denotation quality prove as various and complex as data capturing and producing itself. A few measures have been recalled. Many more exist. They come in infinite numbers to the inventive mind. Some may differ only slightly in purpose, or subtly in their methods. Always, however, their constitutive alchemy will alter (for better or for worse) the meaning of their results, that is, the meaning of what they are meant to account for.

Controling the denotation quality of geographic information summons and combines mathematical tools of all kinds. The juxtaposition of statistical, geometrical, analytical, and other methods may occasionally look like do-it-yourself (or all-made-by-yourself-from-scratch) machinery. "With a little bit more rope and a few extra hairsprings it should be made to work." The expected construct however is no small contraption: it is more of the *bricole* kind, a war-engine akin to the catapult, coherent, well-balanced, finely adjusted, made to break resistances – in the present case, the resistance of the world to being represented. Quite a paradoxical way of cultivating the world's ever-distant beauty, which mobilizes passion and knowledge, and which maps and data not only bring closer, but also intensify, if only by their being in their turn set there in the world.

11.6. References

[ABB 94] ABBAS I., Base de données vectorielles et erreur cartographique. Problèmes posés par le contrôle ponctuel. Une méthode alternative fondée sur la distance de Hausdorff : le contrôle linéaire, 1994, PhD Thesis, IGN, Paris 7 University.

[ALL 91] ALLEN J.F., "Time and time again: the many ways to represent time", *International Journal of Intelligent Systems*, 1991, vol 6, number 4, p 341–55.

[ALT 95] ALT H. and GODAU M., "Computing the Fréchet distance between two polygonal curves", *International Journal of Computational Geometry & Applications*, 1995, vol 5, number 1–2, p 75–91.

[BAR 01] BARILLOT X., HANGOUËT J.F. and KADRI-DAHMANI H., "Généralisation de l'algorithme de 'Douglas et Peucker' pour des applications cartographiques", *Bulletin du Comité Français de Cartographie*, 2001, number 169–170, p 42–51.

[BEL 01] BEL HADJ ALI A., Qualité géométrique des entités géographiques surfaciques. Application à l'appariement et définition d'une typologie des écarts géométriques, 2001, PhD Thesis, IGN, Marne-la-Vallée University.
ftp://ftp.ign.fr/ign/COGIT/THESES/BEL_HADJ_ALI/these.pdf

[BER 67] BERTIN J., Sémiologie graphique – Diagrammes, réseaux, cartographie, 1967, Paris, Gauthier-Villard/Mouton.

[BER 03] BÉRA R. and CLARAMUNT C., "Relative adjacencies in spatial pseudo-partitions", in Proceedings of COSIT 2003, KUHN W., WORBOYS M.F. and TIMPF S. (eds), Lecture Notes in Computer Science 2825, 003, p 204–20.

[BOL 02] BOLLOBÁS B., Modern Graph Theory, 2002, New York, Springer.

[BON 02] BONIN O., Modèles d'erreurs dans une base de données géographiques et grandes déviations pour des sommes pondérées; application à l'estimation d'erreurs sur un temps de parcours, 2002, PhD Thesis, IGN, Paris 6 University.
ftp://ftp.ign.fr/ign/COGIT/THESES/BONIN/theseBonin.pdf

[DAV 97] DAVID B. and FASQUEL P., "Qualité d'une base de données géographique: concepts et terminologie", Bulletin d'information de l'IGN, 1997, number 67.

[DEV 97] DEVOGELE T., "Processus d'intégration et d'appariement de bases de données géographiques. Application à une base de données routières multi-échelles", 1997, PhD thesis, IGN, Versailles University.
ftp://ftp.ign.fr/ign/COGIT/THESES/DEVOGELE/

[DEV 02] DEVOGELE T., "A new merging process for data integration based on the discrete fréchet distance", in Proceedings of the Symposium on the Theory, Processing and Application of Geospatial Data, 2002, ISPRS Technical Commission IV, Ottawa, p 167–81.

[FAÏ 99] FAÏZ S. O., Systèmes d'informations géographiques: information qualité et data mining, 1999, Tunis, éditions Contributions à la Littérature d'Entreprise.

[FON 03] FONSECA F., DAVIS C. and CÂMARA G., "Bridging ontologies and conceptual schemas in geographic information integration", GeoInformatica, 2003, vol 7, number 4, p 355–78.

[GAM 95] GUPTILL S.C. and MORRISON J.L., Elements of Spatial Data Quality, 1995, Oxford, Pergamon Elsevier Science.

[GRI 89] GRIFFITH D.A., "Distance calculations and errors in geographic databases", in Accuracy of Spatial Databases (NCGIA), Goodchild M.F. and Gopal S. (eds), 1989, Taylor & Francis, p 81–90.

[HAN 98] HANGOUËT J.F., Approche et méthodes pour l'automatisation de la généralisation cartographique; application en bord de ville, 1998, PhD Thesis, IGN, Université de Marne-la-Vallée.
ftp://ftp.ign.fr/ign/COGIT/THESES/HANGOUET/

[HAN 04] HANGOUËT J.F., "Multi-representation: striving for the hyphenation", International Journal of Geographical Information Science, 2004, vol 18, number 4, p 309–26.

[ISA 89] ISAAKS E.I. and SRIVASTAVA R.M., *An Introduction to Applied Geostatistics*, 1989, New York, Oxford University Press.

[ISO 00] INTERNATIONAL STANDARDS ORGANIZATION, Systèmes de management de la qualité. Principes essentiels et vocabulaire, 2000, NF EN ISO 9000.

[ISO 02] INTERNATIONAL STANDARDS ORGANIZATION, Geographic Information, Quality Principles, ISO 19113:2002, 2002, Genève, ISO.

[ISO 03] INTERNATIONAL STANDARDS ORGANIZATION, Geographic Information, Quality Evaluation Procedures, ISO 19114:2003, 2003, ISO, Genève, ISO.

[ISO 03b] INTERNATIONAL STANDARDS ORGANIZATION, Geographic Information, Spatial Schemas, ISO 19107:2003, 2003, Genève, ISO.

[KUH 03] KUHN W. and RAUBAL M., "Implementing semantic reference systems", in *Proceedings of the 6th AGILE Conference on Geographic Information Science*, GOULD M., LAURINI R. and COULONDRE S. (eds), 2003, Lausanne, Presses Polytechniques et Universitaires Romandes, p 63–72.

[MAE 01] MAEDCHE A. and STAAB S., "Comparing Ontologies – Similarity Measures and a Comparison Study", 2001, Internal Report number 408, Institute AIFB, Karlsruhe.

[MÜL 01] MÜLLER J.-C., SCHARLACH H. and JÄGER M., "Der Weg zu einer akustischen Kartographie", *Kartographische Nachrichten*, 2001, vol 51, number 1, p 26–40, Bonn, Kirschbaum Verlag.

[OKA 92] OKABE A., BOOTS B. and SUGIHARA K., *Spatial Tessellations – Concepts and Applications of Voronoi Diagrams*, 1992, John Wiley & Sons.

[OSG 57] OSGOOD C.E., SUCI G.J. and TANNENBAUM P.H., *The Measurement of Meaning*, 1957, Urbana. Illinois, University of Illinois Press.

[OTT 01] OTT T. and SWIACZNY F., *Time-integrative Geographic Information Systems – Management and Analysis of Spatio-Temporal Data*, 2001, New York, Springer.

[PAV 81] PAVLOWITCH P., *L'Homme que l'on croyait*, 1981, Paris, Fayard.

[REG 98] REGNAULD N., Généralisation du bâti: Structure spatiale de type graphe et représentation cartographique, 1998, PhD Thesis, IGN, Provence University. ftp://ftp.ign.fr/ign/COGIT/THESES/REGNAULT/

[ROD 03] RODRIGUEZ M.A. and EGENHOFER M.J., "Determining semantic similarity among entity classes from different ontologies", *IEEE Transactions on Knowledge and Data Engineering*, 2003, vol 15, number 2, p 442–56.

[SCH 93] SCHMITT M. and MATTIOLI J., *Morphologie Mathématique*, 1993, Paris, Masson.

[VAU 97] VAUGLIN F., Modèles statistiques des imprécisions géométriques des objets géographiques linéaires, 1997, PhD Thesis, IGN, Marne-la-Vallée University. ftp://ftp.ign.fr/ign/COGIT/THESES/VAUGLIN/

[ZAH 71] ZAHN C., "Graph-theoretical methods for detecting and describing gestalt clusters", *IEEE Transactions on Computers*, 1971, vol C.20, number 1, p 68–86.

Chapter 12

Communication and Use of Spatial Data Quality Information in GIS

12.1. Introduction

All geospatial data are, at different levels, imprecise, inaccurate, incomplete, and not up-to-date, being an approximation of a reality observed through a specific data model, and measured using technologies offering different precisions. However, depending on the use that will be made of these data, these imperfections can be considered as acceptable for a specific use. The expression "best available data" is often used, as users do not always find data that exactly fit their needs. The best available data at a given time could provide acceptable results conveying the notion of "satisfycing" (Simon [SIM 57]), well known in the field of decision-making. The concept of quality is not then absolute, as a single dataset can be of very good quality for one user, but unusable for another one. For instance, a positioning error of some meters will likely be problematic for cadastral issues, but may not be a problem for a user willing to identify the best road to visit a friend.

In order to evaluate the extent to which data are suitable for a given application, users need to get some information regarding the data characteristics (for example, spatial and temporal coverages, spatial and semantic accuracies, completeness, up-to-dateness). Such information will allow them to choose between (1) trying to *reduce* the residual uncertainty by, for example, adapting the dataset, or by looking for a new dataset that is more appropriate, and (2) *absorbing* this uncertainty, and accepting the potential risks of negative consequences (see Chapter 14). Part of this

---

Chapter written by Rodolphe DEVILLERS and Kate BEARD.

information on data quality focuses on the internal quality as measured by data producers and provided to users through metadata (that is, data about data). Although quality information may be documented by some producers and stored by some commercial software systems (e.g. ESRI ArcCatalog or SMMS for Intergraph Geomedia), but it is not still not utilized in GIS calculations or communicated to users in a complete, contextual and understandable way. For example, a distance measure can be provided with great precision (ArcGIS displays measured distances to six decimal places corresponding to a thousandth of a millimeter) even when not warranted by the data source (generalized satellite imagery with documented accuracy of 250 meters). Similarly, users can calculate the number of fire hydrants located within the limits of a city using an inclusion operator and get a precise but inaccurate result if the estimated data completeness for fire hydrants is 90%.

Thus, there is a need to manage data quality information in order to be able to communicate this information more efficiently to users and, hopefully, to use this information to enhance the results of GIS functions. The scope of data quality communication and use is thus to reduce the risks of misuse of geospatial data.

This chapter presents challenges in terms of data quality management and different models to support this management. It presents different methods of communicating quality information to users and ways in which such information can be used by GIS in order to reduce risks of geospatial data misuse. Finally, it presents a software prototype for managing and communicating data quality information before concluding and presenting some future research perspectives.

12.2. Data quality information management

To supply data quality information to users and for GIS or processing purposes, data quality information should be stored in a structured way. Data quality information, especially that advocated by metadata standards, can describe data at different levels of detail. Some information can be associated with an entire dataset (for example, a topographic dataset including roads, buildings, rivers, etc.), other information can describe a single feature type (for example, only buildings) or even a single feature instance (for example, one specific building) (see Figure 12.1). It is important to be able to manage data quality information at different levels of detail in order to benefit from the richness that this information contains. For example, there is limited value in documenting the spatial accuracy at the dataset level when the dataset is highly heterogeneous. Hunter [HUN 01] identified the granularity of data quality information as one of the major concerns for future research in the field of data quality. He notes that quality generally suffers from overly general representations and provides several examples of metadata such as: spatial accuracy is "variable", "from 100m to 1000m" or even "+/- 1.5m (urban) to +/- 250m (rural)".

Furthermore, Hunter mentions that the documentation of data quality at an aggregate level does not capture knowledge of the spatial variability of data quality, although such information would be useful to users. There is thus a need to document data quality at a finer level of detail. The idea of documenting metadata at the feature level is not new but there is a strong economic disincentive, as this would represent a significant additional cost in metadata acquisition for data producers, as well as an increase in the volume of metadata to manage.

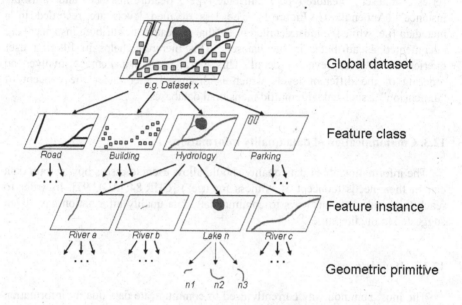

Figure 12.1. *Example of data granularity*

Several authors have worked on the problem of managing data quality information at different levels of detail and proposed different models (for example, [FAI 96; FAI 99; QIU 99; QIU 02; BED 95]). Although different, these hierarchical models often share common levels of detail such as dataset, feature class and feature instance levels. Some hierarchies include the geometric primitives (e.g. point, line, or polyline objects) that make up a feature instance as shown in Figure 12.1.

Some geospatial data producers use these hierarchies to organize their metadata. Metadata of the National Topographic Database (NTDB) of Canada, for example, includes metadata with four levels of detail: dataset, metadata polygons (spatial extents for which some metadata variables are homogeneous), theme (a set of feature classes addressing the same theme), and geometric primitives. Some metadata are provided as a text file while other metadata (e.g. spatial accuracy) are associated with the geometric primitives as attributes. Metadata for the Australian topographic database (1:250,000) are also documented at four levels of detail: dataset, data layer, feature class, and individual feature levels [HUN 01].

The international standard ISO 19115 [ISO 03] suggests a hierarchy to store metadata at different levels of detail. The ISO hierarchy extends Faïz and Qiu and Hunter by taking into consideration the attribute granularity in addition to the geometry (attributes and attribute instances). ISO 19115 levels include: "data series", "dataset", "feature type", "attribute type", "feature instance" and "attribute instance" (Appendix G, Figure 1). The less detailed levels are recorded in a metadata file, while the most detailed ones (that is, feature or attribute instances) are documented as attributes in the datasets. This hierarchy helps in filtering user queries for a given level of detail. This hierarchical management involves an ordering of the different levels which is discussed later under the concept of "dimension" associated with multidimensional databases.

12.3. Communication of data quality information

The information about data quality should allow users to assess how certain data can fit their needs (concept of "fitness for use") [CHR 84; AGU 97]. In order to reach this goal, different ways to communicate data quality information have been suggested in the literature.

12.3.1. *Metadata*

The most common way currently used to communicate data quality information to users is through the diffusion of textual metadata (see Figure 12.2), which include information regarding data quality (see Chapter 10 for more details on metadata). Different organizations publish standards for documenting data quality information and metadata (e.g. ISO 19113 and ISO 19115, Federal Geographic Data Committee (FGDC), European Committee for Standardization (CEN)). One of the objectives of the metadata is to allow users to assess the fitness for use of datasets. However, the usefulness of such metadata remains limited as the content can be difficult to understand by users, even by experts in geographic information [GER 03; TIM 96]. Text-based metadata may thus not be the most efficient way to communicate data quality.

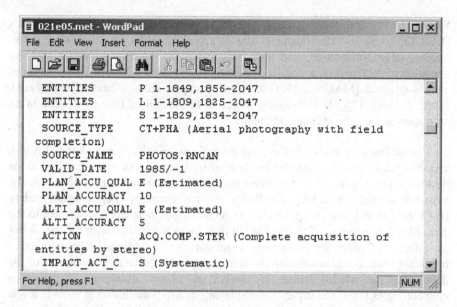

Figure 12.2 *Example of metadata provided with a dataset*

Hunter and Masters [HUN 00a] note that the provision of metadata by data producers is increasingly perceived by users as protection for producers from potential lawsuits rather than as support information for users to assess fitness for use.

12.3.2. *Data quality visualization*

Another approach aimed at communicating quality information relies on visualization techniques. Visualization provides an effective means to present complex information and has thus been proposed as an approach for communicating geospatial data quality to users. Graphical methods can be particularly effective for the evaluation of large volumes of geospatial data [GOO 02] for a number of reasons. The human visual channel is adept at recognizing structure and relationships, graphic and cartographic depictions are a natural and direct expression for spatial structure, and visual comprehension is a fast communication channel capable of supporting high volumes of information. In the spirit of scientific visualization, such methods also provide the opportunity to see the unseeable.

Research on visualizing geospatial data quality has addressed a number of issues, many of which were outlined by the NCGIA research initiative on Visualizing the Quality of Spatial Information [BEA 91]. Much of the initial research focused on the data quality components prescribed by the various metadata standards (for example,

FGDC, CEN) and their association with appropriate visual displays. Bertin's [BER 73] framework for mapping data to visual variables (location, size, value, texture, color, orientation, and shape) provided the guiding principles, but for depiction of data quality, a number of extensions to Bertin's original visual variables were suggested [MAC 92; MCG 93] (for example, color saturation and focus). Fisher [FIS 93; FIS 94] additionally suggests the length of time of display in an animation as a means of encoding quality.

Visualizing data quality is the same problem as visualizing metadata [BEA 98], and thus attention to the structure and linkage of data and metadata is integral to the development of a visualization framework. Effective visualization presumes the existence of useful metadata, specifically relevant measures of error or uncertainty [BEA 99; FIS 99], and an appropriate structure of the metadata with respect to the data. Metadata maintained as text files in only loose association with the data is not an effective format for supporting visualization. As indicated above, another important issue is the association of data quality with different granularities of the data, from a whole dataset to individual object instances. Visualization methodologies should be capable of displaying quality information at various levels of granularity with a clear indication of the data level being described. Explicit association of data quality measures or descriptions with the data in a DataBase Management System (DBMS) can support efficient access to the quality information and assure its association with the data at an appropriate level of detail.

Additional research topics have addressed user interaction issues in quality visualization. User interaction issues involve consideration of graphic design and implementation in combination with an understanding of user cognitive and perceptual abilities. The goal has been to make data quality information available to users in a form that can be easily assimilated and in a context that is useful to their tasks. Quality, as an expression of fitness for use, implies a specific application context and so users need mechanisms to indicate their data quality needs. A number of visualization approaches allow users to specify data quality components and thresholds and view data with respect to these. The concept of a data quality filter [PAR 94] selects and displays only data that meet a user set threshold.

Other user interaction concerns have addressed the question of how to present quality information in combination with the data. MacEachren [MAC 94] described three approaches for associating data with quality descriptions. These included: (1) side by side images, (2) composite images, and (3) sequenced images. Each of these options has associated graphic challenges. Side by side images require the same format for both displays and the presence of some reference objects within both images such that users can visually and mentally co-register the images. Sequenced images also require a common format and scale for both data and quality images. Composite images have the advantage of presenting a single image, but require a graphic encoding of the data that is distinguishable from the graphic encoding of the data quality information. A number of composite schemes make use of different bivariate mapping schemes [BRE 94], but such approaches can rapidly become visually complex and difficult for users to interpret. Robertson [ROB 91] suggested use of terrain representations as a cognitively familiar approach for merging multiple variables in a single image.

There are many dimensions to data quality (see levels of details presented in section 12.2) and many different components as well (named "quality elements" by the standards). A user may be interested in only a single component (for example, positional accuracy) or they may prefer a highly summarized measure over all components. The different needs for quality information combined with variable display approaches for different components suggests the need for a flexible exploratory environment for user interaction with data quality. The Reliability Visualization System (RVIS) package [HOW 96] (see Figure 12.3) and visualization tools added to GRASS (Geographic Resources Analysis Support System) [MIT 95] are two examples of exploratory environments for data quality.

New approaches and measures of data quality continue to be developed and they present new visualization challenges. Recent research on data quality [HUN 97; FIS 98; KYR 99; WIN 02] aims to capture local and finer spatial variation in data quality. However, this finer detail in spatial variation can be more difficult to communicate. Side-by-side and sequenced images are less able to support user association of fine spatial patterns in quality with the data distribution. Additionally, more quality assessment is being supported by Monte Carlo simulation that creates multiple possible realizations of spatial distributions. The visualization challenge becomes one of conveying the uncertainty represented by large sets of possible realizations. A set of realizations available for visualization represents a substantial volume of data for users to effectively assimilate. Kyriakidis et al. [KYR 99] have been investigating visualization approaches for sets of realizations.[1] A number of methods that employ fuzzy-set approaches [LEU 92] generate similar results in that there exist a range of possible values for every location or a range of positions for different features.

----

1 http://www.geog.ucsb.edu/~yoo/nsfviz/

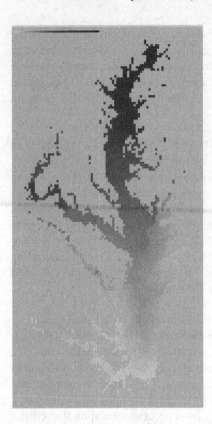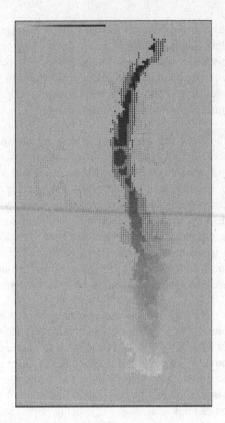

Figure 12.3. *Example of maps displayed by R-VIS software [HOW 96] showing the concept of focus to represent the uncertainty using side-by-side images (initial map on the left and map using the focus to highlight areas with a higher uncertainty on the right) – with authorization of Alan MacEachren, GeoVISTA Center, Penn State*

12.4. Use of quality information

Once the information regarding data quality is stored in a structured database, it is possible to enhance GIS functions with this information. With the exception of some academic work, such capabilities do not yet exist within commercial GIS. For example, several analyses performed by GIS can have significant levels of error, such as counting a number of objects located in a certain area when the completeness is average or measuring distances on a road network when the roads are highly generalized. Information regarding data quality could be used more effectively to refine results, providing users with margins of error, warnings regarding the interpretation of the results, or even restricting the use of some functions when the data would not provide reliable results. Different approaches for employing data quality information to enhance GIS analysis are presented next.

12.4.1. *Warnings and illogical operators*

Reinke and Hunter [REI 02] proposed a theoretical model for the communication of uncertainty, adapted from Gottsegen *et al.* [GOT 99] (see Figure 12.4). In this model, the representation is central to the process of communicating uncertainty. The representation requires feedback between the user and the system, allowing a greater interaction from the user, potentially leading to better communication (that is the feedback loop from Figure 12.4).

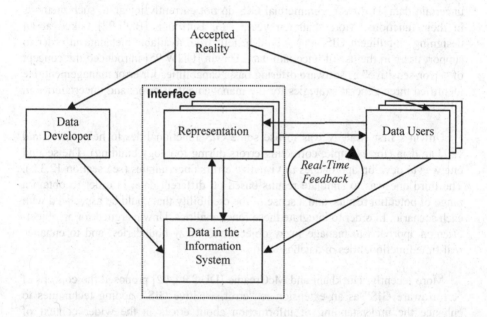

Figure 12.4. *Model to communicate information about spatial database uncertainty to users ([REI 02])*

Based on this model, Reinke and Hunter proposed, using data quality information stored in a database, to send messages to users when they performed operations which were judged illogical based on the quality of the data [HUN 00b]. Rules defined in the system activate the display of messages when some initial conditions are met (for example, if the user measures a distance on the map and the unit of measure is unknown, a warning is then displayed related to the distance measurement). Beard [BEA 89] proposed a similar idea by suggesting to structure *a priori* the data to only allow relevant operations for specific types of information (for example, to avoid mathematical operations on nominal or ordinal data). Functions of this type are now available to some extent in commercial GIS. ArcGIS, for example, does not generate statistics (average, minimum, maximum) for nominal or ordinal data when the appropriate data type is specified. The creation of thematic maps has also been restricted to comply with graphic semiology rules (for example, selecting the representation according to scale of measurement of the data).

12.4.2. *Quality-Aware GIS*

Many works have also addressed the issues of error propagation in GIS, as well as the evaluation of the validity of the results obtained by analyses based on uncertain data. However, commercial GIS do not currently integrate such methods in their functions. More than ten years ago, Burrough [BUR 92] talked about designing "intelligent GIS" that would benefit from available metadata in order to support users in the use of uncertain data. Unwin [UNW 95] introduced the concept of "error-sensitive" as software offering basic capabilities for error management. He identified three general strategies for the management of errors and uncertainties in GIS.

Unwin's first strategy was to use some GIS functionalities to help verify and validate data (for example, correcting errors during topology building). The second one was to develop approaches to visualize errors/uncertainties (see section 12.3.2). The third one was to simulate results based on different data, in order to obtain a range of potential results, and a sense of the credibility that could be associated with each scenario. In order to integrate these functionalities, Unwin argued for an object-oriented approach to manage fuzzy objects and fuzzy boundaries, and to enhance real-time functionalities of databases.

More recently, Duckham and McCreadie [DUC 99, 02] proposed the concept of "error-aware GIS" as an extension to "error-sensitive GIS", adding techniques to enhance the understanding of information about errors in the wider context of decision-making. They highlighted the importance of such systems for commercial GIS, as well as for legal issues (these issues are also discussed in the Chapter 15 of this book). Duckham and McCreadie's "error-aware GIS" is based on an object-oriented structure for data management, combined with artificial intelligence techniques (inductive learning).

In a broader context we could group and extend these approaches under the concept of "quality aware GIS" that integrate different techniques to document, manage, and use quality information to enhance GIS functioning. The term "quality" is broader than the term "error", adding a contextual component in terms of users' requirements for quality. Such systems should be able to:

– integrate quality information into a single database structure (for example, integration of metadata);

– store this information at different levels of detail in relation to the data described (see section 12.2);

– provide tools to support update of quality information (for example, metadata) when changes, such as updates, are made to data in order to keep this information relevant;

– invoke different GIS functions that take into consideration the data quality;

– warn users of potential problems resulting either from the quality of the data used or from the combination of the data of a given quality with some operators (see Hunter and Reinke); and

– provide tools to visualize data quality information.

12.5. Example: multidimensional user manual prototype

Data quality evaluation (fitness for use) can become an extremely complex task, even for experts, because of the heterogeneity of data involved in the decision-making process (spatial and temporal coverages, completeness, logical consistency, date of validity of data, etc.). In this context, based on a study from the juridical field related to geomatics and to the diffusion of digital data, Gervais [GER 03] demonstrated the contribution that an expert in data quality, like an audit, could provide to non-expert users in complex situations. A project named Multidimensional User Manual (MUM) has been developed at Laval University (Quebec) in order to design tools that would help experts in quality to assess the fitness for use in a specific context, and then allow them to provide advice to non-expert users (see [DEV 04, 05] for more details). The prototype of the MUM system is based on the combination of two approaches: (1) a multidimensional database allowing the management of quality information at different levels of detail and (2) an adaptation of management dashboards (balanced scorecards), allowing the communication of quality using indicators embedded into the cartographic interface that supports users in their decision-making process. The prototype was programmed using components from four technologies, (1) *GeoMedia Professional* cartographic components, (2) *SQL Server Analysis Services* for the multidimensional database, (3) *Proclarity* (OLAP client) for navigation within the multidimensional database, and (4) *MS-Access* for the relational database, allowing storage of parameters related to the prototype as well as the user profile.

Multidimensional spatial databases, exploited using tools such as OLAP (On-Line Analytical Processing), allow the structuring of data at different levels of detail and along different dimensions, in order to allow users to navigate within the different levels of details in a fast and intuitive way. The dimensions are not only the traditional spatial (x, y, z) and temporal (t) ones, but are more generally axes of analysis for the users. For the prototype, geometric data from the Canadian National Topographic Database, describing several themes (for example, roads, rivers, buildings, etc.) has been used. Metadata describing data quality, according to the recent international standard ISO 19113 (see Chapter 10), were integrated and structured in the multidimensional database. The data that describe quality have been compared to levels of acceptance, for a given user, in order to categorize the quality into three classes (bad, average, and good). Links were made between metadata and the associated geometric data.

The user interface is composed of four main components (see Figure 12.5): (1) the cartographic component for visualizing the data, as well as visualizing data quality information – bottom right, (2) traditional cartographic tools (for example, zoom in, zoom out, pan) – top center, (3) a dashboard for the display of quality information – on the left – and (4) OLAP tools allowing users to navigate into the information at different levels of detail (for example, move from the overall quality of an entire dataset down to the quality of a single feature class, and then to a single feature instance) – top right.

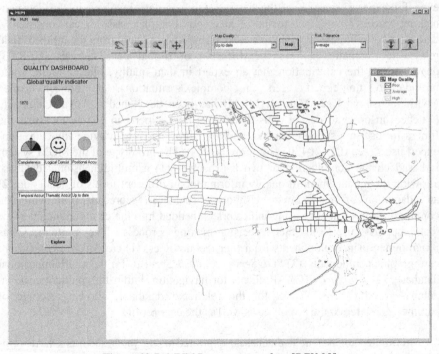

Figure 12.5. *MUM Prototype interface [DEV 05]*

Quality indicators are first selected by the user to display only the relevant indicators. Each indicator is described within a descriptive sheet that indicates how it is calculated, what it represents, and how it can be interpreted.

The user can also navigate into quality information in different ways:

– Navigation into the indicator hierarchy: allows users to navigate in the quality indicators hierarchy (for example, from the overall quality down to the horizontal spatial accuracy for instance).

– Navigation into the data hierarchy: allows users to navigate through data levels (for example, from the entire road network down to road instance, to a line segment of a specific road).

– Cartographic visualization of data quality: allows visualizing quality indicators on a map and highlighting the spatial heterogeneity of quality.

These functions represent a sub-sample of the possible tools that can be offered. They are mainly designed to communicate data quality information, but once the data are structured hierarchically in a multidimensional database, it becomes easy for other functions to use this quality information (to estimate the quality of the objects located within a specific *ad hoc* area defined by the user). Such tools can be used by experts to enhance their knowledge of data quality and to provide advice to non-experts regarding use of the data in specific application contexts. The structure could also be useful to geospatial data producers to help them obtain a quick insight on the quality of the data they are producing, for example, in helping to plan update strategies.

12.6. Conclusion

This chapter has presented a review of the problems of communication and use of geospatial data quality information.

The chapter discussed the importance of communicating quality information to geospatial data users. We first looked at the different considerations related to quality information management, that directly support the communication and use of data quality information. We then presented different approaches to communicate this information. The first one involved the distribution of metadata, which include some information about data quality. The second one involved visualization techniques, to provide a visual representation of quality (often on a map). While the first way is often used, the second one has not yet been implemented in commercial GIS, although research on this issue has existed for more than 15 years.

Additionally, the chapter presented different research aimed at using data quality information to enhance GIS functions and limit the risks of misuse. We explored different approaches proposed in the literature and then presented some recent work on integrating these different approaches to support quality informed GIS functions. In order to illustrate these concepts, we presented in section 12.5 the prototype of a system for managing and communicating quality information. This system allows users to navigate into quality information at different levels of detail. The use of quality information by GIS is relatively recent and limited, but has a promising future ...

12.7. References

[AGU 97] AGUMYA A. and HUNTER G.J., "Determining fitness for use of geographic information", *ITC Journal*, 1997, vol 2, number 1, p 109–13.

[BEA 89] BEARD M.K., "Use error: the neglected error component", *Proc. Auto-Carto 9*, 1989, Baltimore, USA, ACSM-ASPRS, p 808–17.

[BEA 91] BEARD M.K., BUTTENFIELD B.P. and CLAPHAM S.B., NCGIA Research Initiative 7: Visualization of Data Quality, Technical Report 91-26, 1991, Santa Barbara, USA, NCGIA.

[BEA 98] BEARD M.K. and SHARMA V., "Multilevel and graphical views of metadata", *Proc. IEEE Forum on Research and Technology Advances in Digital Libraries*, April 1998, IEEE, Santa Barbara, USA, p 256–65.

[BEA 99] BEARD M.K and BUTTENFIELD B., "Detecting and evaluating errors by graphical methods", in *Geographic Information Systems: Principles and Technical Issues*, Longley P., Goodchild M.F., Maguire D. and Rhind D., (eds), 1999, New York, John Wiley & Sons, vol 1, p 219–33.

[BED 95] BEDARD Y. and VALLIERE D., Qualité des données à référence spatiale dans un contexte gouvernemental, Technical Report, 1995, Laval University, Canada.

[BER 73] BERTIN J., *Sémiologie graphique: les diagrammes, les réseaux, les cartes*, 1973, Paris, Éditions Mouton-Gauthier-Villars-Bordas, p 431.

[BRE 94] BREWER C.A., "Color use guidelines for mapping and visualization", in *Visualization in Modern Cartography*, MacEachren A. and Taylor D.R.F., (eds), 1994, Oxford, Elsevier, p 123–48.

[BUR 92] BURROUGH P.A., "Development of intelligent geographical information systems", *International Journal of Geographical Information Systems*, 1992, vol 6, p 1–11.

[CHR 84] CHRISMAN N.R., "The role of quality information in the long term functioning of a geographical information system", *Proc. Auto-Carto 6*, 1984, Ottawa, Canada, p 303–21.

[CIT 04] CITS, Base nationale de données topographiques, http://www.cits.rncan.gc.ca/, last consulted 16 March 2006.

[DEV 04] DEVILLERS R., BEDARD Y. and GERVAIS M., "Indicateurs de qualité pour réduire les risques de mauvaise utilisation des données géospatiales", *Revue Internationale de Géomatique*, 2004, vol 14, number 1, p 35–57.

[DEV 05] DEVILLERS R., BÉDARD Y. and JEANSOULIN R., "Multidimensional management of geospatial data quality information for its dynamic use within Geographical Information Systems", *Photogrammetric Engineering & Remote Sensing*, 2005, vol 71, number 2, p 205–15.

[DUC 99] DUCKHAM M. and MCCREADIE J.E., "An intelligent, distributed, error-aware OOGIS", *Proc. of 1st Int. Symp. on Spatial Data Quality*, 1999, Hong Kong, The Hong Kong Polytechnic University, p 496–506.

[DUC 02] DUCKHAM M. and MCCREADIE J.E., "Error-aware GIS development", in *Spatial Data Quality*, Shi W., Fisher P.F. and Goodchild M.F. (eds), 2002, London, Taylor & Francis, p 63–75.

[FAI 96] FAÏZ S., Modélisation, exploitation et visualisation de l'information qualité dans les bases de données géographiques, 1996, PhD Thesis in Computer Sciences, Paris Sud Uniersity, France.

[FAI 99] FAÏZ S., *Systèmes d'informations géographiques: Information qualité et data mining*, 1999, Éditions C.L.E., Tunis, p 362.

[FIS 93] FISHER P.F., "Visualizing uncertainty in soil maps by animation", *Cartographica*, 1993, vol 30, number 2/3, p 20–27.

[FIS 94] FISHER P.F., "Visualization of the reliability in classified remotely sensed images", *Photogrammetric Engineering and Remote Sensing*, 1994, vol 60, number 7, p 905–10.

[FIS 98] FISHER P.F., "Improved modeling of elevation error with geostatistics", *Geoinformatica*, 1998, vol 2, number 3, p 215–33.

[FIS 99] FISHER P.F., "Models of uncertainty in spatial data", in *Geographic Information Systems*, Longley P, Goodchild M.F., Maguire D. and Rhind D. (eds), 1999, New York, John Wiley & Sons, vol 2, p 191–205.

[GER 03] GERVAIS M., Pertinence d'un manuel d'instructions au sein d'une stratégie de gestion du risque juridique découlant de la fourniture de données géographiques numériques, 2003, PhD Thesis in Geomatics Sciences, Laval University, Canada, p 347.

[GOO 02] GOODCHILD M.F. and CLARKE K., "Data quality in massive datasets", in *Handbook of Massive Datasets*, Abello J., Pardalos P.M. and Resende M.G.C. (eds), 2002, Norwell, USA, Kluwer Academic Publishers, p 643–59.

[GOT 99] GOTTSEGEN J., MONTELLO D. and GOODCHILD M.F., "A comprehensive model of uncertainty in spatial data", *Proc. Spatial Accuracy Assessment, Land Information Uncertainty in Natural Ressources Conference, Québec, Canada*, 1999, Sleeping Bear Press Inc, p 175–82.

[HOW 96] HOWARD D. and MACEACHREN A.M., "Interface design for geographic visualization: tools for representing reliability", *Cartography and Geographic Information Systems*, 1996, vol 23, number 2, p 59–77.

[HUN 00a] HUNTER G.J. and MASTERS E., "What's wrong with data quality information?", *Proc. GIScience 2000*, 2000, Savannah, USA, University of California Regents, p 201–03.

[HUN 00b] HUNTER G.J. and REINKE K.J., "Adapting spatial databases to reduce information misuse through illogical operations", *Proc. Spatial Accuracy Assessment, Land Information Uncertainty in Natural Ressources Conference*, 2000, Amsterdam, The Netherlands, Delft University Press, p 313–19.

[HUN 01] HUNTER G.J., "Spatial data quality revisited", *Proc. GeoInfo 2001*, 2001, Rio de Janeiro, Brazil, p 1–7.

[ISO 03] ISO TC/211, "ISO 19115 – Geographic Information – Metadata", 2003.

[KYR 99] KYRIAKIDIS P.C., SHORTRIDGE A.M. and GOODCHILD M.F., "Geostatistics for conflation and accuracy assessment of digital elevation models", *International Journal of Geographical Information Science*, 1999, vol 13, number 7, p 677–708.

[LEU 92] LEUNG Y., GOODCHILD M.F. and LIN C.C., "Visualization of fuzzy scenes and probability fields", *Proc. 5th Int. Symposium on Spatial Data Handling*, 1992, vol 2, p 480–90.

[MAC 92] MACEACHREN A.M., "Visualizing uncertain information", *Cartographic Perspectives*, 1992, vol 13, p 10–19.

[MAC 94] MACEACHREN A.M., *Some Truth with Maps: A Primer on Symbolization and Design*, 1994, Washington DC, USA, Association of American Geographers.

[MAC 93] MACEACHREN A.M., HOWARD D., VON WYSS M., ASKOV D. and TAORMINO T., "Visualizing the health of Chesapeake Bay: an uncertain endeavor" *Proc. GIS/LIS '93*, November 1993, Minneapolis, USA, ACSM-ASPRS-URISA-AM/FM, p 449–58.

[MCG 93] MCGRANAGHAN M., "A cartographic view of data quality", *Cartographica*, 1993, vol 30, number 2/3, p 8–19.

[MIT 95] MITASOVA H., MITAS L., BROWN W.M., GERDES D.P., KOSINOVSKY I. and BAKER T., "Modeling spatially and temporally distributed phenomena: new methods and tools for GRASS GIS", *International Journal of Geographical Information Systems*, 1995, vol 9, number 4, p 433–46.

[PAR 94] PARADIS J. and BEARD M.K., "Visualization of spatial data quality for the decision maker: a data quality filter", *URISA Journal*, 1994, vol 6, number 2, p 25–34.

[QIU 99] QIU J. and HUNTER G.J., "Managing data quality information", *Proc. Int. Symp. on Spatial Data Quality*, 1999, Hong Kong, The Hong Kong Polytechnic University, p 384–95.

[QIU 02] QIU J. and HUNTER G.J., "A GIS with the capacity for managing data quality information", in *Spatial Data Quality*, Shi W., Goodchild M.F. and Fisher P.F. (eds), 2002, London, Taylor & Francis, p 230–50.

[REI 02] REINKE K.J. and HUNTER G.J., "A theory for communicating uncertainty in spatial databases" in *Spatial Data Quality*, Shi W., Goodchild M.F. and Fisher P.F. (eds), 2002, London, Taylor & Francis, p 77–101.

[ROB 91] ROBERTSON P.K., "A methodology for choosing data representations", *IEEE Computer Graphics and Applications*, 1991, vol 11, number 3, p 56–67.

[SIM 57] SIMON H.A., *Models of Man*, 1957, New York, John Wiley & Sons, p 279.

[TIM 96] TIMPF S., RAUBAL M. and KUHN W., "Experiences with metadata", *Proc. Spatial Data Handling 96, Advances in GIS Research II*, 1996, Delft, The Netherlands, Taylor & Francis, p 12B.31–12B.43.

[UNW 95] UNWIN D.J., "Geographical information systems and the problem of error and uncertainty", *Progress in Human Geography*, 1995, vol 19, number 4, p 549–58.

[WIN 02] WINDHOLZ T.K., BEARD M.K. and GOODCHILD M.F., "Data quality: a model for resolvable objects", in *Spatial Data Quality*, Shi W., Goodchild M.F. and Fisher P.F. (eds), 2002, London, Taylor & Francis, p 116–26.

Chapter 13

External Quality Evaluation of Geographical Applications: An Ontological Approach

13.1. Introduction

Geographical data come from different sources and it seems easy to combine them, using their common spatial reference, for analysis and decision-making. However, we may wonder what the value is of a decision based on data whose (internal) quality is poorly understood in the context of the decision. This is the problem of "fitness for use" or "external quality" of an application, and it is important to help the decision-maker to evaluate this external quality. There is, presently, no formal tool to help the user to select data in agreement with his quality requirements [CHR 82]. According to Hunter [HUN 01], users would like to have technical facilities to take the quality information into account and to determine what quality and relevance the results coming from the use of these data and models will have. This is currently only carried out in a very limited way by qualified experts, and such functionality generally does not exist in commercial software packages. Frank [FRA 98] suggests that the description of the quality of the data must be independent of the method of production; the users should be able to use procedures that predict quantitatively the quality of the result of the analyses. Hence, there is a need for better evaluation of external quality.

The objective of this chapter is to present an approach that makes it possible to evaluate the external quality of a geographical application. We first define the concepts of quality and ontology. We then present a method known as the

Chapter written by Bérengère VASSEUR, Robert JEANSOULIN, Rodolphe DEVILLERS and Andrew FRANK.

"ontological" approach, which allows one to model, evaluate, and improve external quality of geographical information. Later, we illustrate our method through a practical example to compare the user's needs and the data quality. We conclude this chapter by identifying topics for future research.

13.2. Quality and ontology

Before any qualitative or quantitative assessment of the data quality and of its influence on decision can be made, the ontological status of quality with respect to geographic data must be clarified.

13.2.1. *Quality definition and external quality*

Two definitions of the quality of the data can be identified in the literature (see Chapter 2 for more details). The first connects the quality of the data to the internal characteristics of the data, starting with the methods of data production (data acquisition, data models, etc.). It represents the difference between the produced data and "perfect" data (that is, "ideal"). This definition is often called "internal quality" [AAL 98; AAL 99; DAS 03]. The second definition follows the concept of "fitness for use" [JUR 74; VER 99; CHR 83], connecting the level of adequacy existing between the characteristics of the data and the user's needs for various aspects (for example, spatial and temporal coverage, precision, completeness, up-to-dateness, etc.). This definition, often called "external quality", thus relates to the aptitude of the data to meet the implicit or explicit needs of the user. This chapter focuses on the evaluation of external quality.

Coming from the field of the industrial production, the concept of external quality was introduced in the field of numerical geographical information in 1982 by the U.S. National Committee for Digital Cartographic Data Standards (NCDCDS), and reported in 1987 [MOE 87]. At present, the definition of the quality retained by ISO corresponds to the concept of external quality. It is defined as being the totality of characteristics of a product that bears on its ability to satisfy stated and implied needs.

Although the concept of fitness for use was adopted more than 20 years ago, there has been almost no evolution since then, and almost no development of methods to evaluate the fitness for use of data and models selected for a specified geographical application [VER 99]. The evaluation of whether the data and models are adequate remains in the hands of the end-user, where the evaluation is usually done in an intuitive way, based on the experience of the user, on the information received from an expert of the field of application, or on some information available on the characteristics of the data files or models. The catalogs and metadata which provide information about the "capabilities" of data (e.g. Web service capabilities: see Chapter 10) can help the user to select data among an available subset which is supposed to better fit their interest (time, zone, thematic focus). Research in visualization of uncertainty (see Chapter 12) will eventually help the user in evaluating the fitness for use. More recently, research aimed at establishing more formal approaches for the definition of the fitness of use. Agumya and Hunter [AGU 98] propose to divide these approaches into two categories. The first, based on standards ("standard-based"), compares intrinsic uncertainty of the data with acceptable levels of uncertainty. The second, based on a study of risk ("risk-based"), evaluates the potential impact of uncertain data on the decisions that the user wants to take.

1 – "*Standard-based*"

Frank [FRA 98] presents a "meta-model" that makes it possible to amalgamate the points of view of the producers and users of data [TIM 96]. He criticizes the utility of the currently diffused metadata and demands descriptions of quality more independent of the methods of production, operational and quantitative.

Vasseur *et al.* [VAS 03] present an approach using ontologies to formalize, first, the characteristics of the data as well as, secondly, the user's needs. Ontologies provide two comparable models, facilitating comparison between these two ontologies (that is, quality increases with the similarity). This approach is presented in detail in this chapter.

2 – "*Risk-based*"

Agumya and Hunter [AGU 98] explore the problem from a different viewpoint. They analyze the potential risks *a posteriori* following the use of data for a certain application. Based on the work from the field of risk management, they define various stages in the process: identification of the risk (what can go wrong and why), analysis of risk (probabilities and consequences), evaluation of risk (what is the acceptable risk), exposure to the risk (what is the real level of risk), estimate of the risk (comparison between the acceptable level of risk and the exposure to the risk), and response to the risk (how the risk can be controlled). This approach implies a quantification of the risks, and it may prove to be difficult.

De Bruin *et al.* [DEB 01] propose an approach, based on the concept of "value of control" that allows the comparison of various alternative datasets for an application. This approach, using the technique of decision trees, requires a quantification of the accuracy of the data and an estimate of the costs of an incorrect decision based on these data. It enables, by using a quantitative estimate of the risks, the selection of the data file, limiting the negative consequences in accordance with the decision-making. However, the quantification of the risk is difficult, especially while taking into account uncertainty.

13.2.2. *Complexity of external quality evaluation*

The evaluation of external quality is a comparison between the data characteristics and expectations of the user and is often made in an intuitive manner [GRU 04]. This evaluation is more complex than it appears at first sight. The concept of external quality, related to the ability to satisfy the user's needs, is a dynamic process, fuzzy and complex. The user may modify, in a converging process, his requirements in agreement with the present data, or conversely seek new data to respond to his problem. That is close to the definition of the quality of the field of industrial production of goods, where quality aims to satisfy the customer in a step of continual improvement [JUR 74]. The best quality is obtained when the difference between the quality of the data and the needs is minimal. If a user requires, for example, recent data, one has to deal with questions such as: where to put the limit between recent and non-recent data? Will the user be satisfied with data older than one week, one month, one year? This limit is not only fuzzy, it is also contextual to each user, and also to each use. The characteristics of data, and of the user needs, should be compared in order to compute a "utility value" of the data. The utility constitutes a quantitative measure of external quality, coming from the comparison between expectations of the user and the available data [FRA 04]. That implies having a common framework reference of comparison and it is often necessary to carry out translations and standardizations to be able to compare various characteristics.

We define external quality as the degree of similarity existing between the user's needs and the data, expressed in the same reference frame.

13.2.3. *Concept of ontology*

The ontology can help to compare and measure the similarity between the needs and the data, but this requires formalizing the user's needs and the characteristics of the data. The reality of the world and knowledge that we have are recurring questions in philosophy. The word ontology comes from the Greek roots *ontos* and *logos*, ontology is the science of "what is". The philosophers have studied for centuries the existence, knowledge, and description of "to be". In computer science, ontology is defined as the "explicit specification of a conceptualization" [GRU 93], or as "the whole of the definitions of the classes, relations, and other objects of interest (physical or logical) in a field of interest" [NOY 01]. Ontology, thus, makes it possible to share a vocabulary of discourse, a knowledge base, a common reasoning framework of exchangeable, and comprehensible information, even for non-specialists in the field. There is not only one way to model a field of knowledge, but several alternatives that depend on the application.

In the field of geographical information, Mark defines ontology as "the totality of the concepts, categories, relations and geospatial processes" [MAR 02]. The publication of a geospatial ontology provides a reasoning model that makes it possible to integrate various constraints and to define the semantics and the terminology of a geospatial application [FRA 03]. Hunter describes ontology as a framework to carry out a clear and concise description of terms and concepts that we employ, so that others can interpret what we do [HUN 02]. The literature identifies several levels of abstraction, such as the conceptual representation of a reality (objects, classes, etc.), the logical representation, and the physical representation. For external quality, the various levels of granularity may be, for instance: the description of a problem and data; the conceptual representation of the important features in the problem and in the data; and the specifications of the user needs and of the data. In information sciences, the concept of ontology is very close to that of an object-oriented model associated with a shared vocabulary. We are interested in geographical information and in the quality of this information; the reference to the real world thus intervenes under two points of view: as a reality shared by the user and the producer of data, and as a supposed horizon of a perfect quality. This reference to reality uses the ontology concept in the philosophical sense, as well as concept of ontologies for applications. We need various levels of abstraction: a level of definition of entities useful for the problem considered, a level of representation of these entities, and a translation in an adequate formalism.

13.3. Representation of external quality

13.3.1. *Improvement of quality process*

The notion of quality is related to the notion of user satisfaction, and should participate to a process of continual improvement [JUR 74, ISO 00]. First, the basic goal of the application must be clearly explained and specified. Thereafter, the available elements, compatible with expectations of the user are selected. The user's needs are gradually better achieved while refining the initial conceptualization of the problem. This iterative process converges towards a satisfactory situation for the user, combining the ontology of the problem and of the data. This can be represented in the form of a "quality matrix". Figure 13.1 provides an illustration of this process.

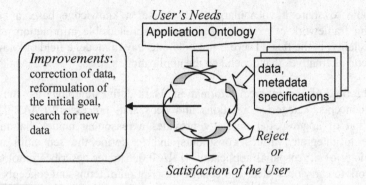

Figure 13.1. *Improvement of quality – adaptation of the "wheel of Deming" [ISO 00]*

13.3.2. *Geosemantical integration*

To evaluate the external quality of the data, one needs to capture the knowledge of the user (ontology of the problem) and of the data (ontology of the product). They must be integrated in an ontology of a common application or reference frame (see Figure 13.2); Brodeur calls it *geosemantical space* [BRO 03]. The various elements in the same reference frame are translated into a common language. Geosemantic integration can also be carried out on a more general level [GUA 98, SOW 98, KOK 01, COM 03]. When the translation is carried out, one can compare two matrices of "expected" and "internal" quality and provide the matrix of "actual" quality (application) that results from the comparison. This representation provides a structure to document, measure, and communicate the quality of the application [VAS 03].

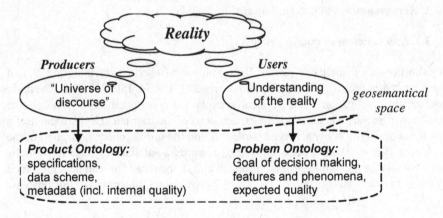

Figure 13.2. *Ontological approach of external quality evaluation*

13.3.3. *Stages allowing external quality evaluation*

The process of improvement of external quality can be modeled in several stages, as illustrated in Figure 13.3.

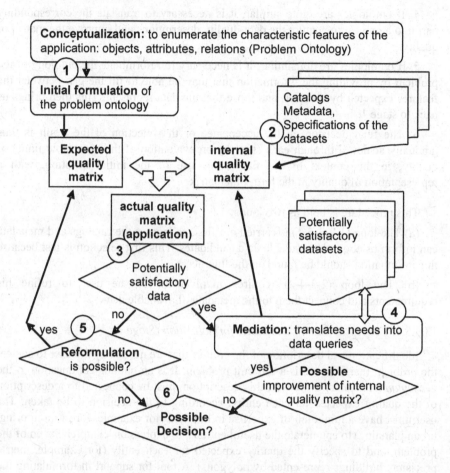

Figure 13.3. *Method of improvement of external quality*

The stages of this process are:

1. Conceptualization: the question and the working hypotheses are inserted in an ontology of the problem; clarifying what is necessary at the beginning, including the requirements in quality, and building the matrix of expected quality.

2. Inspect catalogs of metadata to decide what they contribute in order to understand the quality of the data; the matrix of internal quality is built.

3. The third stage establishes the quality matrix of the application that evaluates whether the data are compatible with the quality expected by the user. If the quality is satisfactory, one carries out the direct translation (stage 4), if not, one seeks another solution in the stage 5.

4. If ontologies are quite similar, it is necessary to translate the corresponding part into a request to acquire the data-quality descriptions and improve the matrix of quality.

5. If ontologies are not similar, it is necessary to reformulate the ontology of the problem by providing the information that may be able to fill the gap between the features expected by the user, and those that the data represent. This then brings us back to stage 1.

6. The final decision on the acceptance or the rejection of the result is thus gradually achieved through each stage by an evaluation of the acceptable limits of quality, in the context of the fitness for use of the initial question, with a representation of quality in the form of a matrix.

This procedure is built as two loops:

(a) The loop (1-2-3-5-1) is carried out first, as long as the catalogs and metadata can inform us about the available data candidates. This first selection is fast because this information should be found on the Internet.

(b) The loop (2-3-4-2) requires involvement of the user to refine his requirements and connect them to the metadata of available data.

13.3.3.1. *Conceptualization and initial formalization (Stage 1)*

Identification and description of the entities that are needed by the user to answer the problem that is posed is a difficult problem. It is an ontological question in the *geosemantical space*. Moreover, this description must be connected to a description of the quality expected for each entity, according to the decision to be taken. The user must have a minimum of expertise in his field (for example, forest monitoring, urban planning) to enumerate the useful entities, to express his comprehension of the problem, and to specify the quality expected for each entity (for example, metric precision, buildings represented by polygons). A tool for support in formulating the initial ontology, or the reformulation (after stage 5), would be useful, but such a tool do not exist yet.

13.3.3.2. *Expected and internal quality matrices (Stage 2)*

One builds two matrices based on the same rows and columns: a first matrix that represents the internal quality of the data (that is, based on quality as measured or documented), and a second matrix describes expected quality (that is, considered to be satisfactory by the user within the framework of its application).

Then one compares these two matrices, cell by cell, using similarity measurements, and the result constitutes a matrix known as the "quality matrix of the application", which identifies the entities for which external quality (that is, difference between data and needs) is achieved.

These three matrices consist of:

(a) Columns that represent entities of the application that are present in "the ontology of the problem" (C_i).

(b) Rows that represent a list of elements of quality, the data elements possible to use (Q_i).

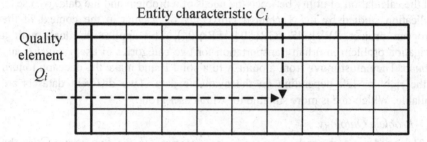

Figure 13.4. *Quality matrix*

13.3.3.2.1. Entities of an application

The geosemantical space can be modeled using entities characteristic (C_i) that can be objects in a given state (for example, a parcel), dynamic phenomena (for example, a flow), sets of dependencies (for example, a network), etc. There is not a single nomenclature for all the entities of an application. Hunter said that there is no good or bad description; there are just different descriptions for different goals, for different people [HUN 02]. One can use vocabularies coming from various techniques of modeling (for example, entity-relationship vs. object-oriented); use the definitions of the UML (Unified Modeling Language) [UML 03], which has the advantage of taking into account temporal dimension of the processes and events in certain geospatial applications. One can be interested in the description of a state in a given moment or, for example, in the description of a continuous process (for example, erosion) [VAS 04].

13.3.3.2.2. Quality elements

The standard ISO 19113 [ISO 02] defines five quantitative elements and 15 sub-elements to help to evaluate internal quality. These elements relate to the parameters of completeness, logical consistency, accuracy, which is subdivided into positional and temporal, and thematic accuracy. In addition, other criteria than those of ISO

19113 are defined in the literature. For example, *temporality* means the difference between the creation date of the data file and the date on which it will be used; *coverage* makes it possible to know the territory and the period that the data describe. *Legitimacy* makes it possible to evaluate the official recognition and the legal range of a data (for example, produced by a recognized organization) and, finally, *accessibility* makes availability of the data possible [BED 95].

13.3.3.3. *Comparison and utility: example of an application in transportation (Stage 3)*

In order to illustrate the evaluation of the comparison from the matrix of quality and the calculation of utility between the needs of a problem and the data, we use an application studied by the Technical University of Vienna in the context of the European project REVIGIS [GRU 04; FRA 04]. The application deals with a navigation problem in urban transportation for two categories of users: *tourists* and *firemen*. They must move from a point A to a point B and make the decision to turn to the right or left depending on the available data. Two different datasets are available. Which one is more suitable for which user group?

– *Problem Ontology:*

The needs and the risk acceptance are different for the two groups. For the firemen, the goal is to reach a destination as soon as possible to answer an emergency. The concept of "time necessary to travel" is very important. Information on accessibility with the street network, the addresses, and the sites of fire hydrants must be of very high quality. For the tourists, the goal is to move downtown to discover the sights, on foot or by public transport, and the "perceived distance" is privileged. It is measured less in physical terms of distance than by availability of information or familiarity with the place.

A matrix of expected quality is made for each user group, associating 16 information elements grouped by topics (for example, names of the streets, numbers of the buildings) and three quality elements (for example, precision, completeness, up-to-dateness). These matrices indicate the quality desired for each type of user and each information element contained in a cell. The quality indicated for a cell indicates the ranges of tolerance. It describes ranges of acceptable values corresponding to the variation tolerated between the ideal value (in the matrix of expected quality) and the real value (in the matrix of available quality) for a given information element. For example, for the attribute "points of interest", the group of tourists will wish to obtain qualities with the following tolerances (Table 13.1): up-to-dateness (≤ 2002), completeness ($\geq 95\%$), and precision (≤ 1 meter). On the other hand, the group of firemen is not dependent on these data to make their way to the fire and thus does not require any minimal quality concerning this thematic element.

Themes "points of interest"	Expected quality	Themes	Expected quality
Tourists		Firemen	
Up-to-dateness	2002	Up-to-dateness	null
Completeness	95%	Completeness	-50%
Precision	1 meter	Precision	null

Table 13.1. *Expected quality for the "points of interest"*

– *Product Ontology:*

Two sets of metadatabases are made available to the two groups and they must decide which dataset to acquire: the numerical "Multi-Purpose Map" (MPM) and the "City Map of Vienna" (CMV). These matrices of available data quality use the same thematic elements as the matrices of expected quality. For example, the attribute "points of interest" is given in the CMV with the following qualities (Table 13.2): up-to-dateness (1999), completeness (90%), and precision (0.5 meter). These thematic elements are not present in the database MPM and the corresponding boxes of the matrix are null.

Themes "points of interests"	Actual Quality (MPM)	Actual Quality (CMV)
Up-to-dateness	*Null*	1999
Completeness	*Null*	90%
Up-to-dateness	*Null*	0.5 m.

Table 13.2. *Available Quality for the characteristic "points of interest"*

– *Comparison of the expected and available data quality:*

The matrices of expected and available data quality are established in the same geosemantical reference frame. It is thus possible to carry out the calculation of utility. It is carried out through several stages of comparison, standardization, and aggregation.

The comparison measures the difference between the expected value and the actual value for each thematic element of similar cells. The description of the thematic elements must be in the same unit of measure (for example, day vs. month, meter vs. cm). In the application of CMV, the values of the matrix of quality of the application were translated to percentage values (or 0 and 1), instead of having values expressed in units of percentage, measures, and year.

Aggregation gives a synthetic vision of quality. It is contextual with the data and the user group and very sensitive to the methodology being used. The rules related to the operations of aggregation, the weight of the various values of the matrix and the list of various exceptions must be explained carefully. For instance, ISO 19114 [ISO 03a] describes several data aggregation operators for data quality information: Aggregated Data Quality Results (ADQR). In the above application, a weighted average of the values of the matrix cells of quality of the application was computed. The results give the utility of the dataset for each user group (Table 13.3), (for details of the computation see [GRU 04] and [FRA 04])

Utility	MPM	CMV
Tourists	82%	39%
Firemen	44%	75%

Table 13.3. *Utility for each user group*

We note that the numerical dataset MPM is more useful (that is, of higher external quality) for the group of tourists (82%), while the CMV dataset is more interesting for the firemen (75%).

13.4. Discussion and conclusion

The methodology presented in this section makes it possible to compare available and expected data quality. The process of comparing quality is evolutionary and converges towards a satisfactory situation for the user, combining the ontology of the problem and that of the data. External quality is related to the calculation of utility that comes from the comparison between expected quality and the internal quality of the data available. We use two matrices of quality, representing the ontology of the problem and the ontology of the product, with the same geosemantical reference. It was then possible, in the present example, to evaluate the external quality of geographical information.

This approach is innovative and bring new ways to evaluate external data quality, but it still has some limitations:

– It is assumed that information on the data and their quality is accessible for each theme to be compared. This is usually done at a general level for the theme as a whole, as it is presented in the example of the application of Vienna where quality relates to the themes.

– The user must be able to identify the important entities of his application and to specify the expected quality, with a tolerance level for each theme. That is contextual to the user, with the application and each theme. The most difficult question is to clarify the two ontologies and to formalize them in the same geosemantical space. A "library" of quality matrices, built on cases already investigated by experts, may guide the user in his selection of the datasets which would better fit his expected quality. The idea is to identify themes that are present in several applications and to model their treatment.

– The three dimensions, space, theme, and time are essential for the analysis and understanding of the changes occurring in dynamic geographical problems. The space models often neglect temporal dimension; in theory, ontologists take into account objects, relations, states, and processes. In practice, many of the models are represented by classes with attributes representing a state with certain relations, leaving events and processes for later [KUH 01]. Hence, it appears that an ontology integrating the temporal dimension in the field of the geographical information systems is a research priority [FRA 03].

Standards, such as ISO [ISO 02; ISO 03a; ISO 03b], underline this comment. Standards make it possible to provide a base for better qualifying and quantifying external quality. However, the standards come from discussions and negotiations between institutions (cartographic agencies or software publishers) but only provide a partial vision of reality. The ISO quality standards partly cover the description of a state at a given time, but do not take into account the description of process and events. In order to study all the aspects of the external quality problem, it is important to describe not only the expected and available data quality, but also the models describing the processes. This step is complex, but important to undertake.

13.5. References

[AAL 98] AALDERS H. and MORRISON J., "Spatial Data Quality for GIS", in *Geographic Information Research: Trans-Atlantic Perspectives*, Craglia M. and Onsrud H. (eds), 1998, London/Bristol, Taylor & Francis, p 463–75.

[AAL 99] AALDERS H., "The Registration of Quality in a GIS", *Proceedings of the 1st International Symposium on Spatial Data Quality*, 18– 20 July 1999, Hong Kong, p 23–32.

[AGU 98] AGUMYA A. and HUNTER G.J., "Fitness for use: reducing the impact of geographic information uncertainty", *Proceedings of the URISA 98 Conference*, 18–22 July 1998, Charlotte, USA, p 245–54.

[BED 95] BEDARD Y. and VALLIERES D., Qualité des données à référence spatiale dans un contexte gouvernemental, Technical report, 1995, Laval University, Canada.

[BRO 03] BRODEUR J., BÉDARD Y., EDWARDS G. and MOULIN B., "Revisiting the Concept of Geospatial Data Interoperability within the Scope of Human Communication Process", *Transactions in GIS*, 2003, vol 7, number 2, p 243–65.

[CHR 82] CHRISMAN, N.R., "A Theory of Cartographic Error and Its Measurement in Digital Databases", *Proceedings of Auto-Carto 5*, 1982, Crystal City, p159–68.

[CHR 83] CHRISMAN N.R., "The Role of Quality Information in the Long-Term Functioning of a Geographical Information System", *Proceedings of Auto Carto 6*, 1983, Ottawa, Canada, vol 2, p 303–21.

[COM 03] COMBER A., FISHER P. and WADSWORTH R., "A Semantic Statistical Approach for Identifying Change From Ontologically Divers Land Cover Data", *Proceedings of the 6th AGILE Conference*, 24–26 April 2003, Lyon, France, p 123–31.

[DAS 03] DASSONVILLE L., VAUGLIN F., JAKOBSSON A. and LUZET C., "Quality Management, Data Quality and Users, Metadata for Geographical Information", in *Spatial Data Quality*, Shi W., Fisher P.F. and Goodchild M.F. (eds), London, Taylor & Francis, p 202–15.

[DEB 01] DE BRUIN S., BREGT A. and VAN DE VEN M., "Assessing Fitness for Use: The Expected Value of Spatial Datasets", *International Journal of Geographical Information Science*, 2001, vol 15, number 5, p 457–71.

[FRA 98] FRANK A.U., "Metamodels for Data Quality Description", in *Data Quality in Geographic Information. From Error to Uncertainty*, Jeansoulin R. and Goodchild M.F., (eds), 1998, Paris, Hermès Editions, p 15–29.

[FRA 03] FRANK A.U., "A Linguistically Justified Proposal for a Spatio-Temporal Ontology", http://www.scs.leeds.ac.uk/brandon/cosit03ontology/position_papers/Frank.doc24-2003.

[FRA 04] FRANK A.U., GRUM E. and VASSEUR B., "Procedure to Select the Best Dataset for a Task", *Proceedings of the Third International Conference on Geographic Information Science*, Egenhofer M.J., Miller H., and Freksa C. (eds), 20–23 October 2004, University of Maryland, USA.

[GRU 93] GRUBER T.R., "A Translation Approach to Portable Ontology Specifications", *Knowledge Acquisition*, 1993, vol 5, number 2, p 199–220.

[GRU 04] GRUM E., VASSEUR B., "How to Select the Best Dataset for a Task", *Proceedings of the International Symposium on Spatial Data Quality*, 15–17 April 2004, vol 28b GeoInfo Series, Bruck an der Leitha, Austria, p 197–206.

[GUA 98] GUARINO N., "Formal Ontology and Information System", in *Formal Ontology in Information Systems*, Guarino N. (ed), 1998, Amsterdam, IOS Press, p 3–15.

[HUN 01] HUNTER G.J., "Keynote Address: Spatial Data Quality Revisited", *Proceedings of the 3rd Brazilian Geo-Information Workshop (Geo 2001)*, 4–5 October 2001, Rio de Janeiro, Brazil.

[HUN 02] HUNTER G.J., "Understanding Semantics and Ontologies: They're Quite Simple Really – If You Know What I Mean!", *Transactions in GIS*, 2002, vol 6, number 2, p. 83–87.

[ISO 00] ISO, ISO 9000 – Quality management systems, International Organization for Standardization (ISO), 2000.

[ISO 02] ISO/TC 211, 19113 Geographic information – Quality principles, International Organization for Standardization (ISO), 2002.

[ISO 03a] ISO/TC 211, 19114 Geographic information – Quality evaluation procedures, International Organization for Standardization (ISO), 2003.

[ISO 03b] ISO/TC 211, 19115 Geographic information – Metadata, International Organization for Standardization (ISO), 2003.

[JUR 74] JURAN J.M., GRYNA F.M.J. and BINGHAM R.S., *Quality Control Handbook*, 1974, New York, McGraw-Hill.

[KOK 01] KOKLA M. and KAVOURAS M., "Fusion of Top-level and Geographical Domain Ontologies Based on Context Formation and Complementarity", *International Journal of Geographical Information Science*, 2001, vol 15, number 7, p 679–87.

[KUH 01] KUHN W., "Ontologies in Support of Activities in Geographical Space", *International Journal of Geographical Information Science*, 2001, vol 15, number 7, p 613–31.

[MAR 02] MARK D., EGENHOFER M., HIRTLE S. and SMITH B., "Ontological Foundations for Geographic Information Science", *UCGIS Emerging Research Theme*, 2002. http://www.ucgis.org

[MOE 87] MOELLERING H. (ed), "A Draft Proposed Standard for Digital Cartographic Data", National Committee for Digital Cartographic Standards, American Congress on Surveying and Mapping Report #8, 1987.

[NOY 01] NOY N., MCGUINESS D., Ontology Development 101, A Guide to Creating Your First Ontology, Stanford Knowledge Systems Laboratory Technical Report KSL-01-05 and Stanford Medical Informatics Technical Report SMI-2001-0880, Stanford University, USA, http://protege.stanford.edu/publications/ontology_development/ontology101.html, 2001.

[SOW 98] SOWA J.F., *Knowledge Representation: Logical, Philosophical and Computational Foundations*, 1998, Pacific Grove, USA, Brooks & Cole.

[TIM 96] TIMPF S., RAUBAL M. and KUHN W. "Experiences with Metadata", *Proceedings of the 7th Int. Symposium on Spatial Data Handling, SDH'96*, 12–16 August 1996, Delft, The Netherlands, p12B.31–12B.43.

[UML 03] UML 2.0, Unified Modeling Language formalism, adopted by OMG http://www.omg.org/docs/ptc/03-09-15.pdf

[VAS 03] VASSEUR B., DEVILLERS R. and JEANSOULIN R., "Ontological Approach of the Fitness for Use of Geospatial Datasets", *Proceedings of the 6th AGILE Conference*, 24–26 April 2003, Lyon, France, p 497–504.

[VAS 04] VASSEUR B., VAN DE VLAG D., STEIN A., JEANSOULIN R. and DILO A., "Spatio-Temporal Ontology for Defining the Quality of an Application", *Proceedings of the International Symposium on Spatial Data Quality*, 15–17 April 2004, vol 28b GeoInfo Series, Bruck an der Leitha, Austria, p 67–81.

[VER 99] VEREGIN H., "Data Quality Parameters", in *Geographical Information Systems*, Longley P.A., Goodchild M.F., Maguire D.J. and Rhind D.W. (eds), 1999, New York, John Wiley & Sons, p 177–89.

Chapter 14

A Case Study in the Use of Risk Management to Assess Decision Quality

14.1. Introduction

14.1.1. *Information quality and decision-making*

Over the past 20 years the subject of quality has entered almost every aspect of our daily lives. For example, parents talk about spending quality time with their children, companies spend large amounts of money putting total quality management procedures into place, universities undergo teaching and research quality assessment exercises, and professional quality management associations and journals have now emerged. The topic of geographic information quality was itself at the forefront of this trend when formal procedures for defining and documenting quality were first developed by some of our leading researchers in the early 1980s.

Quality has now become a key issue for both consumers and creators of geographic information. In today's world, policy-makers, managers, scientists and many other members of the community are making growing use of this form of information. Clearly, the increase in the amounts of data being processed and the use of novice-friendly software has helped contribute to a more information-rich society. However, the increased use of geographic information by non-experts has also increased the likelihood of its misuse and misinterpretation, which in turn can lead to doubt or uncertainty about the quality of the decisions that are being taken with it [FRA 98].

Chapter written by Gary J. HUNTER and Sytze DE BRUIN.

Uncertainty in geographic information clearly has the potential to expose users to undesirable consequences as a result of making incorrect decisions. Therefore, any attempt to cope with uncertainty should have as its objectives: (1) to be able to express the uncertainty associated with an information product in meaningful terms and measures; (2) to develop processes that allow informed choices to be made when faced with competing information products; and (3) to devise mechanisms for reducing any residual uncertainty in an information product to a level where it would not alter a decision taken with it.

Of course, reservations about geographic information quality and uncertainty have not just suddenly arisen, and over the past two decades both the clients and creators of geographic information have become aware of the importance of overcoming the problems associated with this issue. Without doubt, the major cause of their uneasiness is the possibility of litigation as a result of either (1) damage suffered through the use of their information products by other parties, or (2) damage suffered as a result of their decisions. Accordingly, in this paper we seek to educate readers in the use of risk management as one means of satisfying the three objectives listed above, and thus helping to address the concerns of both geographic information creators and consumers.

14.1.2. *Chapter outline*

After this introduction, in section 14.2 we discuss how we might determine the required decision quality for a particular task, and a generic strategy is presented for assessing the quality of the information to be used for the decision to be taken. To assist with this task, section 14.3 introduces the concept of risk management. We then have to strike a balance between the quality required for our decision and the quality residing in our information, so in section 14.4 we present several options available for making this tradeoff. In section 14.5 we work through an actual case study where we apply decision theory and risk management as a means of helping to choose which of two different accuracy Digital Elevation Models (DEMs) is best suited to calculate the volume of sand required for a very large construction site to raise its surface level to a new elevation.

14.2. Determining the required decision quality

It is now generally agreed that the purpose of describing information quality is to determine its quality (or fitness for use) for a particular application that usually requires some form of decision to be taken [CHR 90]. To help achieve this purpose, we need to change our focus from the current emphasis on considering purely the effect of uncertainty on the quality of information, to one of considering the effect this uncertainty has on the end-use of the information. In other words, instead of asking "How good is our information?", we should concentrate on "How good are our decisions?". Dealing with uncertainty in geographic information can be divided into the following topics: modeling, propagation, communication, fitness for use assessment, and the treatment of any residual uncertainty [AGU 99].

The order of these five elements is logical since our knowledge about any one element comes from our knowledge of those that precede it. For example, assessing the fitness for use of information for a particular decision requires that uncertainty in the information be clearly communicated. However, before presenting it to the consumer in a form that is understood, the uncertainty will need to be quantified and this requires an understanding of its propagation through the various processes employed. In turn, this presumes that appropriate uncertainty models are available for application to the original input data. Assessing fitness for use requires that we establish whether the quality of the information is acceptable for a given application, and this involves answering two questions:

– What is the uncertainty in the information?
– What degree of uncertainty is acceptable in the information?

After answering these questions, users should be in a position to determine: (1) whether the information uncertainty is unacceptable, in which case they should proceed by reducing it (for example, by collecting more accurate data); or (2) whether the information uncertainty is acceptable, whereby they need to absorb (accept) any residual uncertainty and proceed with the decision. However, in order to compare the answers to these two questions, the results need to be expressed in meaningful units so that decision-makers can directly equate uncertainty in the information with its effect on the decision to be taken.

Traditionally, we have adopted a standards-based approach to dealing with quality in geographic information, but this has largely been due to the use of map-making standards and quality assurance mechanisms since the maps were invariably the end-product for decision-makers. However, now that we have the ability to take a variety of digital datasets, combine and process them, and then produce complex information outputs, the original map/dataset standards do not easily translate into output quality indicators. Also, since uncertainty in the data is considered to be acceptable if its consequences are acceptable, then the standards approach inherently reflects a certain threshold of acceptability – but the question here is "Who are the standards acceptable to?", particularly when data providers rarely know who is using their data or for what purposes. Finally, the standards-based approach does not provide for estimation of the consequences of uncertainty and it is argued that this is one of its key limitations [AGU 99].

As such, we contend that risk management is a more valuable method of deriving information about the risks confronting decision-makers who use geographical information which contains uncertainty [GRA 87; KAP 97; MEL 93; REJ 92]. According to this approach, for each decision task it is imperative to determine what the uncertainty in the information translates into in terms of the risk associated with the decision. Similarly, acceptable levels of uncertainty must be translated into acceptable risk. Thus, two key problem areas are identified in the process of assessing fitness for use based on risk analysis:

– translating uncertainty in the information into risk in the decision (risk analysis);

– determining the level of risk considered acceptable for a given decision task (risk appraisal).

14.3. Using risk management to assess decision quality

Risk management requires the propagation of data uncertainty into decision uncertainty, which is followed by formal risk scenario identification and analysis to quantify the risk in the decision that can be traced back to uncertainty in the data (Figure 14.1). The definition of risk adopted here is that it is comprised of three parts: (1) one or more scenarios of adverse events associated with unwanted decision outcomes; (2) consideration of the likelihood of these scenarios occurring; and (3) their impacts.

Risk management involves a series of tasks. First, risk identification involves documentation of the possible risk scenarios that might occur due to uncertainty in the information. This is considered the most important step in the process on the grounds that "a risk identified is a risk controlled", and essentially involves determining what may go wrong and how it might happen. Next, risk analysis requires estimating the probabilities and expected consequences for the various risk scenarios. The consequences of an adverse event will differ according to the magnitude of the event and the vulnerability of the elements affected by the event. The degree of data utilization will also have an impact here as some data are considered so important that they have the potential to reverse a particular decision, while other data may only have minor reference value.

The outcome of risk analysis is risk exposure (the estimated risk). Risk exposure is the total amount of risk exposed by the adverse event, and can be considered to be the summation of all the individual risks identified. On the other hand, risk appraisal (the acceptable risk) can be determined by analyzing and choosing the risk associated with the most favorable of the possible combinations of decision quality indicators, namely cost-benefits and risks. A complication with this step is the difficulty in quantifying the intangible benefits of a decision. Next, risk assessment requires comparing risk exposure with the results of risk appraisal. Depending on whether the risk exposure is acceptable or not, the decision-maker must then consider taking an appropriate risk response. Risk response is the final stage in the process and also the ultimate objective of risk management – to help the decision-maker choose a prudent response in advance of a problem.

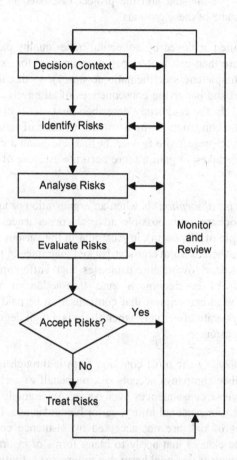

Figure 14.1. *The application of risk management to decision making*

14.4. Dealing with risk in the decision

The possible responses to risk exposure are *risk avoidance, risk reduction, risk retention* and *risk transfer*.

Risk avoidance is basically the "do nothing" option and is selected when a decision-maker is deemed to be highly risk-averse. It would appear to be unreasonable not to proceed with a project because of the risk associated with it, however we are increasingly seeing governments failing to take action on problems because of the perceived political, social, economic or environmental risk. So risk avoidance is certainly not unusual and the project discussed in section 14.5 was actually cancelled on some of these grounds.

Risk reduction is often achieved by collecting better quality data (and we know that this can mean more than just better spatial accuracy – for example, improved logical consistency, completeness or thematic accuracy), so that the uncertainty in the decision is reduced and hence the consequences of an adverse event that might occur are also reduced. In the case study described next, the decision-maker might choose to bring into the construction site a smaller amount of sand than is actually needed, so that when the project site is near its final elevation a detailed survey of the surface can be undertaken to gain a more accurate estimate of the final amount of sand required.

Risk retention (or *risk absorption*) is when an organization or individual chooses to accept the risk associated with possible adverse consequences of using certain information. An example of this occurs in Australia in the various state government land-title offices where there is a government-backed guarantee in place to the effect that customers of the land ownership databases that suffer any harm (usually financial) as a result of conducting a land transaction in which erroneous government data was used can request that compensation be paid to them. Clearly though, this can only operate effectively when there is a high degree of integrity in the data being sold to customers.

Finally, with *risk transfer* the most common form is through insurance where a third party (the insurance company) accepts risk on behalf of a client and makes a payment if certain adverse consequences ever occur. An example of this would be when insurance is taken for personal injury, car or home damage. However, we also know that some types of risk are not accepted by insurance companies, and an example of this are the clauses that apply to many forms of insurance policies and state that property damage or personal harm due to terrorist activities is not covered. With respect to geographic information, it has been possible for a number of years now for data providers to take database insurance policies which protect them against liability and negligence claims by customers in the case where errors in the data cause harm (financial or personal) to users.

14.5. Case study: determining the volume of sand for a construction site[1]

14.5.1 *Case study description*

Early in 2000, plans were made to build a Multimodal Transport Center (MTC) near Nijmegen in the east of the Netherlands. The MTC was proposed as a junction between different transport modes (road, rail, and water). It would lead to reduced CO_2 emissions and simultaneously promote the economic development of the region. The plans included construction of a container port that would require an area of approximately 41.2 hectares to be raised to 15 meters above the Amsterdam Ordnance Datum (NAP), as shown in Figure 14.2. Our case study concerns the expected financial loss due to uncertainty in the calculated volume of sand required to raise the level of the construction site. A more detailed description of the case can be found in [BRU 01].[2]

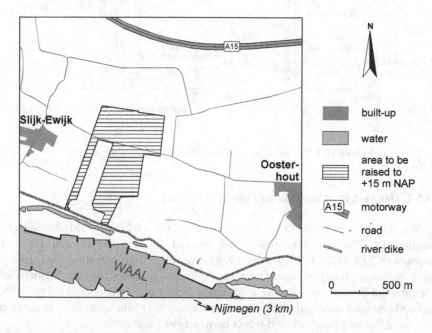

Figure 14.2. *The case study site showing the construction area to be elevated*

---

1 This section is partly based on: De Bruin S., Bregt A. and Van De Ven M., 2001, "Assessing fitness for use: the expected value of spatial data sets", *International Journal of Geographic Information Science*, vol. 15, no. 5, p 457–71. © Taylor & Francis Ltd. We gratefully acknowledge Taylor & Francis for granting permission to use this work.
2 The project was abandoned early in 2003 due to financial, economic, and environmental concerns.

We will consider the situation where a decision-maker in charge of the project is faced with two problems: (1) which DEM should be chosen for volume computations; and (2) what volume of sand should be brought in to the site. In this case study, it is assumed that DEM inaccuracy is the only source of error in the computed volume of sand required for raising the terrain (for example, land subsidence will not be considered). Therefore, computations based on a perfect DEM would result in the exact volume being computed and acquired and, therefore, zero financial loss to the project. However, DEMs will always contain some amount of error due to their approximate nature. By choosing from several alternative DEMs, the decision-maker has partial control over the error process and can therefore manipulate the expected loss rates. Figure 14.3a shows the corresponding decision tree.

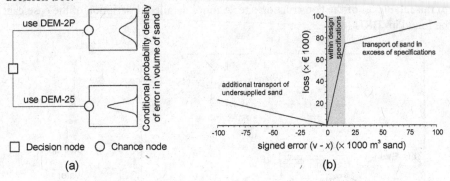

Figure 14.3. *The case study decision tree (a) and loss function (b)*

14.5.2. *Description of the data and model to be used*

We will first examine the fitness for use (quality) of two DEMs having grid spacings of 25m and 2m, which are referred to as DEM-25 and DEM-2P respectively. DEM-25 (dating from 1970) originated from elevation data derived from hardcopy 1:10,000 scale topographic maps produced by the Netherlands Topographic Service (and having approximately one point per hectare). This DEM was updated with surveyed elevations of artificial (built) structures, such as the river dike and other large construction works between 1980 and 1990.

On the other hand, DEM-2P was generated by integrating topographic data with elevation data obtained via photogrammetric image-matching of aerial photography (1999), plus a limited number of terrestrial point elevation measurements. Its average density amounts to approximately 58 points per hectare; however, most of these points have limited accuracy since they were acquired by photogrammetric methods and are certainly not as accurate as those derived directly from ground survey.

Let us assume that any DEM elevations with values lying within ±2cm of the design surface (+15m NAP) are left untouched and require no sand to be placed at their location. The price of sand extracted from a suitable site 2.6 km distant equals €4.55m^{-3} (Alberts, personal communication, 2000). Elevations in excess of +15.02m NAP or below +14.98m NAP would require additional road transport of sand to or from the extraction site at a cost of €0.091m^{-3}km^{-1} (Alberts, personal communication, 2000).

Zero loss would correspond to the event that the determined volume of sand results in a mean elevation of exactly +14.98m NAP, and the resulting asymmetric loss function L(v, x) associated with an error in the determined volume v is shown in Figure 14.3(b) and equation 14.1.

$$L(v,x) = \begin{cases} -0.2366 \cdot (v-x) & \text{if } v - x \leq 0 \\ 4.55 \cdot (v-x) & \text{if } 0 < v - x \leq 16480 \text{ m}^3 \quad [14.1] \\ 74984 + 0.2366 \cdot [(v-x) - 16480] & \text{if } v - x > 16480 \text{ m}^3 \end{cases}$$

In this equation, v is the volume of sand acquired and x is a realization of a random volume X which is bounded by the project extent (41.2ha), the +14.98m NAP surface plane, and the mean DEM-reported elevation with the uncertain mean DEM error over the project area added (or subtracted) to the base surface (which as we know is usually an irregular surface). The figure 0.2366 (€m^{-1}) represents the additional transport cost of undersupplied or oversupplied sand, 16480 (m^3) is the volume of sand falling within the design tolerances (0.04 × 4.12 × 10^5) and 74984 (€) is the cost of that volume.

We assume here that the decision maker wants to minimize expected loss (since any profit – that is, a lowering of cost – is a bonus). This expected loss associated with a particular decision on the acquired volume of sand (v) is calculated by integrating the loss function (Figure 14.3b, equation 14.1) over the probability distribution of the uncertain mean elevation. Typically, this is done by discrete approximation according to equation 14.2:

$$E[L(v,x) \mid \text{DEM}] = \int_{-\infty}^{+\infty} L(v,x) dF(x \mid \text{DEM})$$

$$\approx \sum_{k=1}^{K+1} L(v, \bar{x}_k) \cdot [F(x_k \mid \text{DEM}) - F(x_{k-1} \mid \text{DEM})]$$

[14.2]

In this equation, $F(x \mid \text{DEM})$ is the cumulative distribution function of the uncertain X given the DEM that is used, where x_k, $k = 1, \ldots, K$ are threshold values discretizing the range of variation of x-values and $\overline{x}_k$ is the mean of the interval (x_{k-1}, x_k). Because of the Central Limit Theorem, we can assume X to be normally distributed.

The standard deviation of the mean volumetric error as predicted by block kriging [ISA 89] amounts to $12.24 \times 10^3 \text{ m}^3$ for DEM-2P and $12.67 \times 10^3 \text{ m}^3$ for DEM-25 [BRU 01]. Equation 14.2 was evaluated in a spreadsheet using a polynomial approximation for the cumulative normal probabilities [ROH 81].

14.5.3. *Results*

Figure 14.4 shows the expected loss curves for the two DEMs. Observe that as a result of the asymmetric loss curve and DEM uncertainty, a loss-minimizing decision-maker would decide to acquire approximately $2.0 \times 10^4 \text{ m}^3$ less sand than would be computed on the basis of DEM elevations. The mechanism behind this is clear – it is much cheaper to correct an undersupply of sand than to apply sand in excess of the minimum design specifications. If the DEM-derived amount of sand was acquired then there is a strong chance that too much sand will be applied.

Figure 14.4 also shows that the minima of the expected loss curves for the two DEMs are almost identical. In fact, the difference amounts to a mere €202 [BRU 01], so it can be concluded that the DEMs are about equally suitable for the project. In fact, this highlights the value of the risk management exercise and trying to quantify what effect any uncertainty in the data has upon the final decision. In this case, a project manager could be reasonably assured that DEM error will not be a major contributor to any project losses, and that consideration can be focused on other issues, such as construction time lost due to bad weather, variation in material cost prices, and so on. Finally, Figure 14.4 is useful in showing that a perfectly accurate DEM would reduce the expected loss by approximately €6,000, but the questions that could then be asked are "How much would it cost to create such a DEM?" and "Is it actually possible to create an exact DEM?". We will leave the answers to these questions for readers to consider. Finally, another case study in this area can be found in [BRU 03] regarding the application of decision theory to farm-subsidy fraud detection in the European Union.

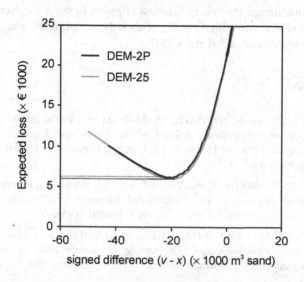

Figure 14.4. *The expected loss associated with departures from the required volume of sand, computed from two DEMs. The dashed lines indicate the difference between the minima of the two curves, and this provides a measure of the fitness of DEM-25 with respect to DEM-2P.*

14.6. Conclusions

In this chapter we have discussed the effects of geographic information quality on decision-making. We have suggested that the starting point for such a discussion is not to ask "How good is our information?", but instead "How good are our decisions?" We believe this is what data users really want to know, since they do not want to take critical decisions based on poor quality information. To achieve this we contend that formal risk management techniques are a valuable means of estimating the likelihood of adverse consequences occurring due to uncertainty in the information used for a particular decision task. So we are advocating a top-down approach to dealing with information uncertainty, instead of the bottom-up method commonly used by the geographic information community.

To demonstrate our approach we have presented a practical case study in the application of risk management that relates to calculating the volume of sand that needs to be brought into a large construction site to raise it to a given elevation – with the obvious risk being one of financial loss if too much material is brought to the site and then needs to be removed.

With respect to future research, there are still several issues that need to be resolved concerning adoption of the risk management approach for assessing geographic information quality, viz.: "Is it cost effective to use it for every decision task, or only those that carry high risk?", "Can it be simplified?" and "Can it be automated and possibly embedded into GIS?".

14.7. References

[AGU 99] AGUMYA A. and HUNTER G.J., "A risk-based approach to assessing fitness for use of geographical information", *Journal of the Urban and Regional Information Systems Association*, 1999, vol 11, number 1, Chicago, Urban and Regional Information Systems Association, p 33–44.

[BRU 01] DE BRUIN S., DREGT A. and VAN DE VEN M., "Assessing fitness for use: the expected value of spatial data sets", *International Journal of Geographical Information Science*, 2001, vol 15, number 5, London, Taylor & Francis, p 45–471.

[BRU 03] DE BRUIN S. and HUNTER G.J., "Making the tradeoff between decision quality and information cost", *Photogrammetric Engineering & Remote Sensing*, 2003, vol 69, number 1, Washington DC, American Society of Photogrammetry and Remote Sensing, p 91–98.

[CHR 90] CHRISMAN N.R., "The error component in spatial data", in *Geographical Information Systems: Principles and Applications*, Maguire D.J., Goodchild M.F. and Rhind D.J. (eds), 1990, London, Longman, vol 1, p 165–74.

[FRA 98] FRANK A.U., "Metamodels for data quality description", in *Data Quality in Geographic Information: From Error to Uncertainty*, Goodchild M.F. and Jeansoulin R. (eds), 1998, Paris, Hermès editions, p 15–29.

[GRA 87] GRATT L.B., "Risk analysis or risk assessment: a proposal for consistent definitions", in *Uncertainty in Risk Assessment, Risk Management and Decision Making*, Covello V.T., Lave L.B., Moghissi A. and Uppuluri V.R.R. (eds), 1987, New York, Plenum Press, p 241–49.

[ISA 89] ISAAKS E.H. and SRIVASTAVA R.M., *Applied Geostatistics*, 1989, New York, Elsevier Science.

[KAP 97] KAPLAN S., "The words of risk analysis", *Risk Analysis*, 1997, vol 17, number 4, London, Blackwell, p 407–17.

[MEL 93] MELCHERS R.E., "Society, tolerable risk and the ALARP principle", *Proceedings of the Conference on Probabilistic Risk and Hazard Assessment*, Melchers R.E. and Stewart M.G. (eds), 1993, Balkema, Rotterdam, p 243–52.

[REJ 92] REJDA G.E., *Principles of Risk Management and Insurance*, 1992, New York, Harper Collins, p 673.

[ROH 81] ROHLF F.J. and SOKAL R.R., *Statistical Tables*, 2nd edn, 1981, New York, Freeman.

Chapter 15

On the Importance of External Data Quality in Civil Law

15.1. Introduction

Geographical information has not escaped from the deep changes that our society has experienced recently. Formerly handled exclusively on analog format, and mostly by expert users, geographical information is now presented more and more in digital format and is disseminated widely on the information highway. The *conversion to digital format* constituted a true revolution in the field of information in general and, specifically, in the field of geographical information. It is now possible to store information within a database, to carry out selective extractions, to combine data of various sources, to produce derived data or to perform various operations or analyses, thanks to database management systems (DBMS) and modern geographical information systems (GISs).

A second revolution, characterizing geographical information, comes from the explosion of application domains, the increase in the number of organizations using geographical information and the wide availability of geographic data on the Internet. In short, we have entered into an era of *mass consumption* of geographic information. The multiplication of digital libraries of geographical data on the Internet, the increased capacity of computers, the ease of loading and transferring data, the growing user-friendliness of geographical information systems and web-mapping applications, and the multiplication of new products intended for the public

Chapter written by Marc GERVAIS.

all contribute to a massive circulation of digital geographical information [WAL 02, CHO 98]. Everywhere in the industrialized world, geographical information is now accessible from multiple sources (free or not), collected by various methods and various producers [MOR 95, AAL 97, AAL 02]. The *democratization process* of geographical information has been confirmed several times in the past by the geomatic community [ROU 91, COT 93, MOR 95, GUP 95, ONS 97, GRA 97, HUN 99].

The two revolutions affecting geographical information (the conversion to digital format and its democratization) have raised several concerns, in particular on the risks of inappropriate use, mistakes and bad interpretations of data or results [AZO 96, HUN 99] by non-experts or unfamiliar users. With new production methods and presentation modes of geographical information, we are likely to see new consumption modes that are as yet incompletely identified and managed.

Several elements suggest a potential increase in the number of disputes over digital geographical information. First, consumers are more and more distant from the original data producer. The loss of contact between these two entities (generated by the Internet for example) constitutes a first source of risk. Users can no longer rely on informal contact to exchange warnings on precise details, particularly concerning the quality and the value of the information presented. Secondly, it is possible to make an analogy between the evolution of the traditional (industrial) economy and the economy of information. Mechanization and the extraordinary rise of the technology brought an increase in the risks of accident and damage to goods and people [BAU 98a]. Regarding databases in general or geographical information products, we currently observe a similar process of mechanization, but with information this time. Technology is now replacing what constituted the knowledge of the cartographer with algorithms and automated processes supported by various specialized software. If the same causes produce the same effects, an increasing risk of disputes resulting from defective information diffusion is to be expected [MON 98].

A probable increase in disputes raises the question of *civil liability*.[1] This is considered by some people as being the most explosive aspect arising from the strong democratization of digital geographical information everywhere in the world.

---

1 Civil liability consists primarily of the duty to repair the damage caused to another when certain conditions are met. Thus, being responsible in civil law consists of answering personally to the pecuniary consequences associated with the behavior which one has in relations with others [PAY 01].

In the United States, in particular, some court decisions caused a chilling effect of tort claims against geographical data producers [OBL 95, PHI 99].

Civil lawsuits can result from defective quality in marketed geographical information. The concept of quality can be divided into two different but complementary, approaches, that is, *internal quality* and *external quality*.[2] Internal quality evaluation measures the degree to which the data acquisition processes adhere to the specifications. External quality evaluation measures the appropriateness of the specifications to the user's needs [DAS 02]. Scientific research undertaken over the last two decades have been aimed especially to evaluate internal quality. The general approach of producers is that the evaluation of external quality or the burden of interpreting the geographical data correctly for a specific use remains the user's responsibility [DAV 97, FRA 98].

This chapter aims to explore the concept of quality according to the civil law (applicable in France, in Quebec and in other places) and to show the importance that producers should attach to external quality evaluation at the time when geographical information is marketed and distributed. To do our analysis, we will address the civil law rules that are applicable to two specific liability regimes, that is circumstances where there may be an *act or fault of another* and circumstances where there may be an *act of a thing*. The result of the analysis indicates that the liability regime does not matter because the court decision will probably be based mainly on the external quality evaluation.

15.2. Applicable general civil liability regimes

There are two principal regimes of liability established by the civil law, extra-contractual liability and contractual liability. Extra-contractual liability is applicable when an event that generates (physical, material or moral) damage occurs between two people without any bond between them (for example, a motor vehicle collision), whereas contractual liability is applicable when the dissatisfaction of one party results from the non-fulfillment of one or several duties enclosed in a contract which links the two parties (for example, to honor a guarantee on a motor vehicle). The two regimes of liability are based on similar concepts [BAU 98a], which are included in the traditional trilogy of fault, prejudice (or damage), and the bond of causality linking the fault and the damage.

---

2 These concepts of internal quality and external quality are more amply discussed in Chapter 3 of this book.

The concept of civil liability is closely related to the concept of duty. In its legal sense, a duty is defined as being the bond of right, existing between two or several parties, in which one party (called a debtor) is held accountable towards another party (called a creditor) to do or not to do something, under the threat of a legal constraint [BAU 98b]. When an individual, by his behavior, transgresses a duty or when a party bond by a contract does not fulfill his obligations, the individual then is at fault. The fault can be seen as the violation of a pre-existing duty, the failure of a pre-existing duty or the derogation from a behavioral standard [PAY 01].

As with databases in general, the applicable regime of liability arising from the marketing of the geographical databases remains uncertain. In fact, if information is considered juridically as being the fruit of an activity, the applicable liability regime would then be the *act or fault of another* regime. When information is presented in a standardized form and is intended for a large audience, information can qualify as a good or product and the liability regime that would then apply is the *act of a thing* regime. In the first regime, the court will consider the producer's behavior, before, at the moment of, and after the marketing of the database. In the second regime, the court will consider more the quality of the product. An attentive examination of the application of the two regimes will reveal the importance of the external quality during the marketing of geographical information.

15.2.1. *Liability regime of an act or fault of another*

Putting into circulation a database is commonly considered as a provision of a service [MAR 97, MON 98]; geographical databases do not escape from this qualification [COT 93]. The juridical situation is indifferent to the format used, analog or digital [PER 96].

The regime of liability for information agencies could be applied to databases [COT 93, DUB 00, LET 01, LET 02, VIV 02]. Under the regime of liability usually called the *charge of defective information*, information will be considered defective when it is inaccurate, incomplete or out-of-date, with these three components becoming the legal standard of the internal quality of information.

Geographical data are obtained by observation in time and space. Because the data come from either primary sources (for example, field surveys) or secondary sources (for example, map scanning or digitizing), the position of the objects in time and space will practically never be in perfect conformity with the real position on the ground [ZHA 02, WAL 02]. The variations, more or less significant, depend on the means implemented by the producer and on the purpose for which data are produced. It is therefore impossible to ensure absolute accuracy, that is, to obtain exact results without errors or uncertainty, and this in spite of the effort made to manufacture geographic information [CHO 98, WAL 02, GER 04].

Geographical information is produced by one or more experts (cartographers, databases designers, etc.) whose decisions, which are often not documented, strongly influence the nature and the number of the imperfections. The decisions are necessarily taken according to one or several implicit or explicit purposes. Throughout the production process, the decisions taken aim to ensure good relative accuracy, that is, an accuracy sufficient to fulfill one or more identified needs. A data file can be perfectly suitable for a previously targeted use and yet be completely inconsistent for a different use. The data are invariably collected and prepared for a particular purpose. Their use in contexts other than that targeted can make these data erroneous or inadequate for a new application.

Thus, taken in their absolute, geographical data are inaccurate, incomplete, and out-of-date in temporal, geometric, and descriptive dimensions [GER 04]. Consequently, the internal quality of geographical data does not conform to the legal standard of internal quality mentioned above. In a strict sense, or in the absolute sense, geographical information would be constantly defective because it is recognized that geographic information is almost always inaccurate, incomplete or out-of-date.

The producer of information is also held to have a general duty of *care* and *diligence*. As regards activities of information, this duty is transformed into a duty of *objectivity* which is divided into a duty of *checking* and duties of *impartiality* and *honesty*. The duty of *objectivity* concerns information in itself (its contents or internal quality) and the diffusion of the information (its presentation and use or external quality) [MAR 97].

For geographical information, the general duty of care and diligence remains difficult to impose for the following reasons. First, the duty of diligence requires that the producers correct erroneous information as soon as they are informed that it is incorrect, a duty that is difficult to apply in regard to databases in general [MON 98]. As geographical data are observational, errors are inevitably present within the database as soon as it is released, and this situation is with the knowledge of the producers.

Secondly, the duty of checking means that the provider of services must diffuse information that is accurate. In a strict sense, accuracy would aim to make the information provided (formally) conform (in time and in extent) to reality [MAR 97], the conformity being appreciated in the absolute, that is, objectively [MON 98]. The determination or evaluation of conformity would also cause difficulties for any databases [LUC 01].

In the case of geographical information, control and checking are carried out, in the majority of cases, by comparing the information with a theoretically ideal model of reality[3] (for example, in France with the IGN "nominal ground" or a data file of higher quality), rather than comparing to the reality itself. As every model of reality is subjective, such evaluation loses its objectivity or absolute character and falls into the sphere of subjectivity or relativity [COT 93].

Thirdly, the evaluation of the conformity of the information provided and compared to reality (by confrontation with the facts) cannot be complete since the checking procedures of the quality of the databases are based on sampling procedures.[4] As the volume of data collected and treated is so substantial, an exhaustive checking on the ground which implies that each data item are checked individually would be extremely expensive. A certain number of erroneous data remains statistically inevitable. Thus, several errors could not be considered as a fault.[5]

Fourthly, the internal consistency check is carried out on the basis of a process of comparison between the database and the specifications fixed by the producer himself. The question which arises then is to ascertain if there are specifications specific to a careful and diligent or reasonable producer. We ignore the criteria that will allow us to establish that such or such a specification emanates from the reasonable man and not from others. This question remains, for the moment, unanswered. The subjectivity that is present during the production of information compounds the subjectivity within checking procedures and results.

In short, producers would be expected to check what at first appears unverifiable. The apparent material and technical impossibility of checking had meant that the duties of the information suppliers were traditionally recognized as being a duty of

---

3 The reader can refer to Chapter 11 on the evaluation of the data quality.
4 The reader can refer to Chapters 10 and 11 on the sampling procedures.
5 The fact of providing erroneous information (non-exhaustive) is not regarded as a fault in itself [COT 93, HAU 00]. Errors, such as mistakes, inadvertencies, relaxations of attention, unfortunate behaviors, are statistically inevitable.

means[6] [COT 93, MON 98] as opposed to a duty of result. Producers have to be careful and diligent in implementing all the means necessary to produce data, but without guaranteeing the results or the accuracy of the provided information. In this context, the fault becomes an error of control, such that it would not have been made by an informed person, placed in the same circumstances.

Objective liability cannot, however, be avoided. Even if the checking remains, in practice, unrealizable for all the data, one should expect it if this were theoretically possible [MAR 97]. The case of the *conium* (poisonous mushroom)[7] and the defective aeronautical chart,[8] which have been frequently quoted [PER 96, MAR 97, MON 98, PHI 99], are convincing examples of the application of an objective liability regime. To meet the completeness criteria, the technical character, by assumption being verifiable [MAR 97] and the data-processing character of the databases [LUC 01] would cause more severe duties for the producers. But completeness would not be necessary in all circumstances, except when information could be legitimately expected to be so by the user [MON 98] or when the service would relate to information that can be normally listed, for example, codes, companies' lists or directories.

The appreciation of the legal standard of quality, that is, the criteria of accuracy, completeness, and currency, raises thorny problems for geographical databases. First, the standard, taken in a strict sense, remains an unattainable goal. It is technically impossible for a producer to reach perfection in all circumstances [CHO 98, WAL 02]. Secondly, its technical and data-processing character can cause an objective appreciation leading to the duty to carry out a complete checking of information, which the volume of data would make painful and expensive.

---

6 The "duty of means" requires that the producer (the debtor) implements all the reasonable means, without however ensuring the user (the creditor) of a particular result, whereas the "duty of result" requires that the producer reach a precise and given result expected by the user. The distinction between the duty of result and the duty of means is often based on the criterion of hazard [MAR 97, LET 02].

7 *Winter v GP Putnam's Sons*, 938 F.2d 1033 (9th Cir. 1991). A French publisher was condemned for having marketed a practical guide for mushroom gathering which was found to have led to the poisoning and the death of a reader [MON 98].

8 *Brocklesby v United States of America*, 767 F.2d. 1288 (9th Cir. 1985). A publisher of navigational charts was found to have marketed a defective chart of an airplane landing approach which did not show an existing geographical object – a mountain. This omission caused the airplane to crash with the death of all of its passengers.

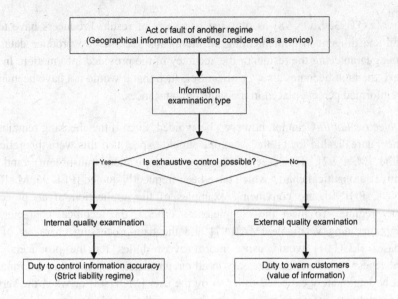

Figure 15.1. *Act or fault of another regime model*

Faced with such a situation, the debate should be about the examination of external quality of information. In that case, the courts will not examine the duty of accuracy with regard to the information content (internal quality), but will consider the duty of accuracy regarding the presentation of the information (external quality) [MAR 97]. The measurement of quality would then become a result of the perception of the customer with regard to the transmitted information [FER 00].

If the producer can not guarantee the information internal quality, he would have a duty to transmit warnings to the user. Customers must be able to appreciate the value of this information [MON 98]. The producer will also be under a duty to draw the customer's attention to the potential misuse or inappropriate interpretations [HAU 00]. In a legal dispute, the factors likely to be taken into account by the court include the type of information, the importance of the information, and the use to which is likely to be made. A fault would not necessarily arise from the presence of erroneous data, but rather from a failure to draw attention to the fact that information is of doubtful validity. The creation process leading to the information would not be questioned and would not give rise to a fault, instead we would speak of *faulty dissemination* [MAR 97]. It would be thus the liability of the supplier to transmit cautionary statements or warnings alerting the customer or specifying the degree of uncertainty in the information [MON 98, HAU 00].

As such, the relative character of accuracy with regard to geographical data forces the producer to attach an importance to external quality. It is difficult to detach the legal assessment of quality from the purposes for which the data are used. For example, when understanding the completeness of the database, missing information would be considered as a fault only when the customer can reasonably expect such information to be absent. It is only when considering the intended purpose that one will be able to determine if the missing information deprives the activity of all its sense.

15.2.2. *Liability regime of an act of a thing*

More and more, the standardization of the information and its mass marketing raise the possible existence of true informational products. The principal consequence would be to modify the nature of the regime of liability. Thus, if information is considered as a product, the regime of liability of an act of a thing will apply.

Just like the European Directive on defective product liability by which it is inspired[9] [MIN 93, BAU 98a, LAM 00, VEZ 03], the regime of liability that applies to the professional manufacturers and salesmen and which is in force in Quebec, applies to tangible goods and movable property (repetition of applies). For some, the software and the databases are excluded from the application of this regime because of their immaterial aspect [MAR 97, LET 01, LUC 01], materiality being an essential condition.

Navigational charts are likely to be subjected to the application of the Directive and, by analogy, to the Quebec regime of liability for professional manufacturers and salesmen. In the support of this thesis, arguments are based on the presence of certain conclusive American court decisions[10] and on the technical nature of navigational charts,[11] their sometimes dangerous character,[12] the specific modes of

---

9 "Directive européenne du 25 juillet 1985 au rapprochement des dispositions législatives, réglementaires et administratives des États membres en matière de responsabilité du fait des produits défectueux (J.O.C.E., 7 août 1985, L 210/29)".
10 *Times Mirror Co v Sisk*, 593 P.2d. 924 (Ariz.1978); *Aetna Casualty & Surety Co.* v *Jeppesen & Co.*, 642 F.2d 339 (1981); *Brocklesby v United States of America*, 767 F.2d. 1288 (9th Cir. 1985); *Winter v GP Putnam's Sons*, 938 F.2d 1033 (9th Cir. 1991). In all these lawsuits, the manufacturers were held strictly liable. This regime would be applicable for all products introducing a risk in their operation, their use or their inaccuracy [PER 96]. The European Directive had borrowed some of its elements from American law [EDW 98], making these American court decisions potentially applicable in France as well as in Quebec through similar borrowing.
11 As opposed to writings of an intellectual nature.
12 In the above-mentioned *Brocklesby* lawsuit, the court declared that a navigational chart (or a product) can be regarded as a *dangerous* product insofar as it is used not for simple reading,

use[13] (different from the other conventional publications) of these and on the possibility of comparing navigational charts to a compass.[14] For all these reasons, the application of the Directive or regime of liability to professional manufacturers and salesmen would be possible.[15]

For one author [PER 96], the distinction between navigational charts (comparable to a compass or a guide) and conventional publications does not appear to be tenable since there are many publications which describe processes or recipes (how-to books). It would be logical to also impose on them the objective liability regime but in reality, according to this author, this would not be the case. A navigational chart as much resembles a book of recipes as it does a compass. The author admits, however, that works of a practical character are more apt to cause physical injuries, thus justifying the imposition of a more severe liability on the producers or manufacturers. Just like the dividing line between a product and a service, the dividing line between functional and non-functional objects is not easy to establish [MON 98].

If a geographical database is considered to be such a product, it would be subject to liability for fault through an objective liability[16] [MAR 97] within which the victim would not have to prove a fault in the actions of the producer,[17] but rather the *defect of the thing*. This regime would fit in the current of extension of liabilities based on the risk created [LET 01].

The producer would then be compared to a salesman and would be held to the same legal duties. However, many of these duties are based on the product's

---

but for practical use. In the above-mentioned *Winter* case, the encyclopaedia had been compared to charts for aerial navigation, that is to say with charts of natural elements which tend to encourage a dangerous activity [MON 98].

13 The charts would not be intended only for simple reading, but to be used as a guide.

14 The compasses would be subjected to this regime and, as the function of the charts is the same as a compass, they should be subjected to the same treatment.

15 Westerdijk [1992, quoted by MON 98] wrote a thesis that tried to show that databases were excluded from the field of the products to which the European Directive applies. The author accepts the idea that software, just like navigational charts, can be regarded as a technical instrument with functional matter and, on this basis, like defective products.

16 The victim is required to prove the defect of the product, but the victim does not have to seek the origin of the defect. The examination of the production process is without effect [MON 98]. Contrary to the service where the appreciation of the fault can rest on the checking processes at the time of the data collecting or processing [COT 93], the important criteria remain the degree of safety and the context of use.

17 One of these problems arises from the plurality of actors implied in the manufacture of databases, making it difficult for the victim to make a liability claim [MON 98]. The liability regime for information agencies could not therefore be used as a model [LUC 01].

expected use or on the anticipated results, that is, external quality. First, the *legal guarantee of quality* is closely connected to the concept of defect. The defect can be understood on two levels; first of all, according to a subjective approach (*in concreto*), that is, the good must be used for the uses envisaged and expressed by the purchaser at the time of the transaction,[18] then according to an objective approach (*in abstracto*), that is, the good is evaluated compared to its normal use or, in other words, according to the use for which it is normally intended (*fitness for purpose*).

Marketing without the control of possible uses can cause ignorance of the uses to which the product is intended and, in consequence, difficulty in the identification and the qualification of a normal use. It is accepted that if the purchaser uses the good improperly and damages result from such abuse, he will not be able to call upon the legal guarantee of quality to support his lawsuit against the producer. If he does not precisely follow the directions for use and if that omission can reasonably explain the defect, the court will thus assume there to be a bond of causality between the omission and the defect [JOB 75]. However, it will be up to the salesman to prove this improper use [LAM 00]. In the absence of previously identified uses, it will be difficult to distinguish between a normal use and an improper use.[19]

According to one author [EDW 98], a defect is not just any imperfection of the good: it is limited to the imperfections that prevent the use of the good. The concept of defect rests on the assumption that any product is intended for a certain use, just as in the case for geographical information. Some defects are not a defect within the meaning of the guarantee. A defect prevents expected uses; the two concepts (defect and use) are conceptually different, but inseparable. The purchaser must establish proof of the bond between the cause (the defect) and the effect (inability to use the product). A defect is not reprehensible itself, except if it prevents normal use. The existence of a defect is thus defined by its result. The result (the use) would be ensured by the prohibition of a cause (the defect) causing the undesirable state (the deficit of use). In conclusion, it can be claimed that the guarantee of quality rests on the protection of the anticipated result and not on the prohibition of the causes that cause a failure. The level of quality of the product would remain indifferent in this respect.[20]

---

18 Before a sale is concluded, if the purchaser explains to the salesman the expected use of the product, the purchaser would be protected because these considerations played a significant role during the formation of the contract.

19 How can a purchaser, being unaware of all processes of design and manufacture of the product, be able to decide between what is an improper use and what is not? How can a producer who is completely unaware of the subsequent uses of the product be able to warn the purchaser that he crosses this limit?

20 The fact that two goods of the same species have different levels of quality does not necessarily constitute a defect with regard to the good which is of worse quality. The reference in the evaluation of defect thus does not come only from one comparison between

Secondly, the producer would have a *duty of delivery*, which implies the delivery of a good conforming to what was agreed, in its quality, its quantity, and its identity. Conformity is understood according to two aspects, which are initially based on the characteristics of the good in question which is described via specifications or the reference frame of conformity which generally corresponds to the metadata.[21] Conformity is analyzed functionally and the stress would be laid on the aptitude of the thing to fulfill its expected use [LET 01]. A fault with regard to the duty of delivery would be a defect in design which would become a defect in conformity when the delivered thing, whilst correct, would not be appropriate to the use it had been ordered for or was expected. The expected characteristics would be performance, availability, and the response time of the product, all of which are elements that are difficult to evaluate without knowing the uses to which the product will be put.

The duty of delivery forces the producer to give the purchaser the intellectual accessories, such as maintenance and operation or instruction manuals [LET 01]. The goal of these handbooks is to provide adequate guidance to the user. If the product does not fix the targeted uses, it will be difficult for the producer to identify the contents of this user manual.

Thirdly, the producer will have a *duty of safety*, which can be divided into two complementary duties. The first duty is the implicit duty to inform the purchaser of the dangers hidden and inherent in the good, when the producer knows them or is supposed to know them. The second one is the duty to inform the purchaser of the precautions to take in order to use the purchased good correctly and to fully benefit from it, for example, by giving instructions to the consumer [BAU 98a]. In fact, the concept of good faith would impose on the manufacturer the duty to inform the consumer about how to use the thing. In the absence of such instructions, the producer could be liable. Here, the duty of information transforms into the duty of safety which constitutes an independent duty connected to the use of a good [LEF 98]. The duty of safety would exist even in the presence of an improper use if the professional could reasonably anticipate the danger but not if no such anticipation were possible [LET 02]. If the producer doesn't target specific uses for his product, it will be difficult for him to act adequately in relation to this duty of safety.

Producers should be concerned with their duty of safety since digital geographical information reveals a significant degree of danger or risk. Indeed, its relative value, and even sometimes its mitigated reliability, raises an unquestionable

---

two goods of the same species, but rather compared to the use expected by a reasonable purchaser.
21 The reader can refer to Chapter 10 on the metadata.

potential to harm, that has been confirmed in the past. In the United States, some indexed court decisions confirm that geographical information products constitute a source of potential harm, sometimes with heavy consequences. There are many examples, such as the bad positioning of underground telephones or electric conduits having led to cutting these,[22] an error in relation to the depth of the seabed making the recovery of a wreck impossible,[23] a measurement of erroneous distance distorting the calculated costs of transport,[24] an error of scale on an approach chart causing the death of all the airplane passengers,[25] a defective approach chart not showing an existing geographical object – a mountain – which caused the death of the airplane passengers,[26] a map that lead to a shipwreck,[27] the collision of a ship where an underwater obstacle was not indicated on the chart,[28] an imprecise map having led to an unreasonable delay after a fire emergency call,[29] an incomplete map having led a hunter to kill an animal in a forbidden area,[30] an incomplete map having contributed to the death by hypothermia of a lost skier,[31] and a badly interpreted map having led an owner to build a building in a flood zone.[32] In each case, the damage was sufficient to initiate legal proceedings. As the dangerous character of geographical information has already been recognized with regard to conventional charts being used for navigation, we believe that digital geographic information must have the same character because the possibilities of mistakes are huge.

The risky character is not limited to the information itself, but also to its mode of marketing. The marketing of geographical information still remains in a state of gross immaturity, especially in the face of mass consumption [LON 01]. Several commercial databases are currently available and accessible to all people who pay the purchase price. By disseminating geographical information to non-expert users, by providing only small indications of error's sources connected to the data-gathering and by not carrying out any control on the subsequent use, producers place themselves in a potentially critical situation [GRA 97], even an explosive one [CHO 98].

---

22 *Bell Canada c Québec (Ville)*, [1996] A.Q. 172 (C.S.); *Excavations Nadeau & Fils. c Hydro-Québec*, [1997] A.Q. 1972 (C.S.).

23 *Fraser Burrard Diving Ltd. v Lamina Drydock Co. Ltd.*, [1995] B.C.J. 1830 (B.-C.S.C.).

24 *Côté c Consolidated Bathurst*, [1990] A.Q. 64 (Qué. C.A.)

25 *Aetna Casualty & Surety Co. v Jeppesen & Co.*, 642 F.2d 339 (1981).

26 *Brocklesby v United States of America*, 767 F.2d. 1288 (9th Cir. 1985); *Times Mirror Co v Sisk*, 593 P.2d. 924 (Ariz.1978).

27 *Algoma Central and Hudson Bay Railway Co. v Manitoba Pool Elevators Ltd*, [1966] S.C.R. 359; *Warwick Shipping Ltd. v Canada* [1983] C.F. 807 (C.A.).

28 *Iron Ore Transport Co. v Canada*, [1960] Ex. C.R. 448.

29 *Bayus v Coquitlam (City)*, [1993] B.C.C.S. 1751; *Bell v Winnipeg (City)*, [1993] M.J. 256.

30 *R v Rogue River Outfitters Ltd.* [1996] Y.J. 137 (Y.T.C.).

31 *Rudko v Canada*, [1983] C.F. 915 (C.A.).

32 *Sea Farm Canada v Denton*, [1991] B.C.J. 2317 (B.-C.S.C.).

According to jurisprudence, the strength of the duty to inform, which is imposed on a manufacturer, may vary according to the circumstances, the nature of the product, and the knowledge or the quality of the purchaser. The duty to inform would be particularly strong when there is danger [BAU 98a, LET 01], a characteristic present in geographical information. Relevant circumstances would include the presentation of the product, the reasonably expected use of the product,[33] and the timing when putting the product into circulation [MON 98]. The presentation of a product would cover information on the product and, in particular, instructions in the form of a user manual or an explanatory note. The manufacturer's duty to inform would also be extended to a professional purchaser who does not have an expertise, specific to geographic information [LET 01].

In short, in the case of the "act of a thing" liability regime, the duties of the geographical information producers (guarantee of quality, duty of delivery, and duty of safety) are closely related to external quality or to the product's expected use. In addition, the relative accuracy and the *potentially* dangerous nature of geographical information increase the duty of the producer to inform the purchaser appropriately and to suitably instruct the consumer about the application of the product or the database for a specific use.

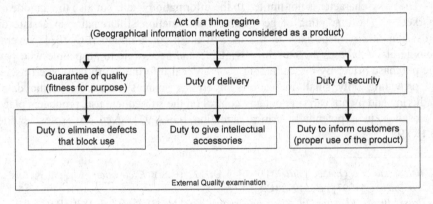

Figure 15.2. *Act of a thing regime model*

---

33 The legitimacy of the expectation will be more contestable if the product were not used in accordance with its purpose or with the regulations recommended for its employment by the producer, but still, it is necessary that these regulations exist.

15.3. Conclusion

Many geographical databases are marketed without any provision for use (application-free), that is, without a targeted or recommended specific use. The producers leave the user to decide if the data are appropriate to the expected uses. In other words, the assumption is posed that the user has the specialized knowledge that is necessary to handle information adequately. However, geographical information is characterized by its technicality and its complexity [COT 93, BEN 95, ZHA 02] and requires a sharpened expertise to understand the limits and the possibilities in terms of treatments and analyses. Application-free marketing would be adequate only to professionals of the same specialty, which constitutes a small minority in a mass market.

Currently, the efforts invested in data quality evaluation relate mainly to the internal elements (specifications) and remain limited, even non-existent, for the external elements (uses). However, the entire production process is strongly tinted by a sequence of decisions taken to specifically target one or several uses. A data quality evaluation should be carried out according to these targeted uses [HAR 02, REI 02]. Even if the data quality is known, it does not necessarily become possible or relevant to use it for all kinds of application [CAS 02]. The targeted use of the data would unquestionably affect the content and the strength of the assumed duties [MON 98].

In conclusion, it does not matter that a geographical database is considered as a service or as a product. The examination of the various legal duties to which the producer would be held shows very clearly the dominant position of external quality. The legal duties also raise the need to transmit *information relating to the use* or to provide instructions or a user manual that can clearly guide a user toward the expected results.

15.4. References

[AAL 97] AALDERS H. and MORRISON J., "Spatial Data Quality for GIS", *Geographic Information Research: Trans-Atlantic Perspectives*, 1997, GRAGLIA M. and ONSRUD H. J., London, Taylor & Francis, p 463–75.

[AAL 02] AALDERS H.J.G.L., "The Registration of Quality in a GIS", in *Spatial Data Quality*, 2002, SHI W., FISHER P. F. and GOODCHILD M. F. (eds), New York, Taylor & Francis, p 186–99.

[AZO 96] AZOUZI M. and MERMINOD B., "Qualité des données spatiales", *Mensuration, Photogrammétrie, Génie Rural*, 1996, vol 12, p 645–49.

[BAU 98a] BAUDOUIN J.-L. and DESLAURIERS P., *La Responsabilité Civile*, Cowansville, Quebec, 1998, Les Éditions Yvon Blais, p 1684.

[BAU 98b] BAUDOUIN J.-L. and JOBIN P.-G., *Les Obligations*, 1998, Cowansville, Les Éditions Yvon Blais, p 1217.

[BEN 95] BENSOUSSAN A., *Les SIG et le droit*, 1995, Paris, Éditions Hermès, p 249.

[CAS 02] CASPARY W. and JOOS G., "Statistical Quality Control of Geodata", in *Spatial Data Quality*, SHI W., FISHER P. F. and GOODCHILD M. F. (eds), 2002, New York, Taylor & Francis, p 106–15.

[CHO 98] CHO G., *Geographic Information Systems and the Law, Mapping the Legal Frontiers*, 1998, London, John Wiley & Sons, p 358.

[COT 93] COTE R., JOLIVET C., LEBEL G.-A. and BEAULIEU B., *La Géomatique, Ses Enjeux Juridiques*, 1993, Québec, Publications du Québec, p 252.

[DAS 02] DASSONVILLE L., VAUGLIN F., JAKOBSSON A. and LUZET C., "Quality Management, Data Quality and Users, Metadata for Geographical Information", in *Spatial Data Quality*, 2002, SHI W., FISHER P. F. and GOODCHILD M. F. (eds), New York, Taylor & Francis, p 202–15.

[DAV 97] DAVID B. and FASQUEL P., "Qualité d'une base de données géographiques: concepts et terminologie", *Bulletin d'information de l'IGN, Rapport 67*, 1997, Saint-Mandé, Institut Géographique National, p 51.

[DUB 00] DUBUISSON B., "Introduction", *La Responsabilité Civile Liée à L'information et au Conseil*, 2000, DUBUISSON B. and JADOUL P., 2000, Brussels, Publications des Facultés universitaires Saint-Louis, p 9–13.

[EDW 98] EDWARDS J., *La Garantie de Qualité du Vendeur en Droit Québécois*, 1998, Montréal, Wilson & Lafleur, p 375.

[FER 00] FERNANDEZ A., *Les Nouveaux Tableaux de Bord des Décideurs: Le Projet Décisionnel dans sa Totalité*, 2000, Paris, Les Éditions d'Organisation, p 448.

[FRA 98] FRANK A.U., "Metamodels for Data Quality Description", in *Data Quality in Geographic Information: From Error to Uncertainty*, JEANSOULIN R. and GOODCHILD M.F., 1998, Paris, Éditions Hermès, Paris, p 15–29.

[GER 04] GERVAIS M., Pertinence d'un manuel d'instructions au sein d'une stratégie de gestion du risque juridique découlant de la fourniture de données géographiques numériques, Québec and France, 2004, Joint PhD thesis, Laval University and Marne-la-Vallée University, p 344.

[GRA 97] GRAHAM S.J., "Products Liability in GIS: Present Complexions and Future Directions", *GIS LAW*, 1997, vol 4, number 1, p 12–16.

[GUP 95] GUPTILL S.C. and MORRISON J.-L., "Look ahead", in *Elements of Spatial Data Quality*, GUPTILL S. C. and MORRISON J.-L., 1995, New York, Elsevier Science, p 189–98.

[HAR 02] HARVEY F., "Visualizing data quality through interactive metadata browsing in a VR environment", in *Virtual Reality in Geography*, FISHER P.F. and UNWIN D., 2002, New York, Taylor & Francis, p 332–40.

[HAU 00] HAUMONT F., "L'information environnementale: la responsabilité des pouvoirs public", *La Responsabilité Civile Liée à L'information et au Conseil*, 2000, Bruxelles, Publications des Facultés universitaires Saint-Louis, p 103–46.

[HUN 99] HUNTER G.J., "Managing Uncertainty in GIS", in *Geographical Information Systems: Principles, Techniques, Applications, and Management*, 2nd edn, 1999, vol 2, GOODCHILD, M. F., LONGLEY P. A., MAGUIRE D. J. and RHIND D. W., London, John Wiley & Sons, p 633–41.

[JOB 75] JOBIN P.-G., *Les Contrats de Distribution de Biens Techniques*, 1975, Québec, Les Presses de l'Université Laval, p 303.

[LAM 00] LAMONTAGNE D.-C. and LAROCHELLE B., *Les Principaux Contrats: La Vente, le Louage, la Société et le Mandat*, 2000, Cowansville, Les Éditions Yvon Blais, p 731.

[LEF 98] LEFEBVRE B., *La Bonne Foi dans la Formation des Contrats*, 1998, Cowansville, Les Éditions Yvon Blais, p 304.

[LET 01] LE TOURNEAU P., *Responsabilité des Vendeurs et Fabricants*, 2001, Paris, Les Éditions Dalloz, p 242.

[LET 02] LE TOURNEAU P., *Contrats Informatiques et Electroniques*, 2002, Paris, Les Éditions Dalloz, p 268.

[LET 02] LE TOURNEAU P. and CADIET L., *Droit de la Responsabilité et des Contrats*, 2002, Paris, Les Éditions Dalloz, p 1540.

[LON 01] LONGLEY P.A., GOODCHILD M.F., MAGUIRE D.J. and RHIND D.W., *Geographic Information Systems and Science*, 2001, London, John Wiley & Sons, p 454.

[LUC 01] LUCAS A., "Informatique et droit des obligations", *Droit de l'informatique et de l'Internet*, 2001, Paris, Thémis Droit privé, Presses Universitaires de France, p 441–588,.

[MAR 97] MARINO L., *Responsabilité Civile, Activité d'information et Médias*, 1997, Aix-en-Provence, Presses Universitaires d'Aix-Marseille et Economica, p 380.

[MIN 93] MINISTERE DE LA JUSTICE, *Commentaires du Ministre de la Justice*, 1993, Québec, Les Publications du Québec, p 2253.

[MON 98] MONTERO E., *La Responsabilité Civile du Fait des Bases de Données*, 1998, Namur, Belgique, Les Presses Universitaires de Namur, p 564.

[MOR 95] MORRISON J.L., "Spatial data quality", in *Elements of Spatial Data Quality*, 1995, GUPTILL S. C. and MORRISON J.-L., New York, Elsevier Science, p 1–12.

[OBL 95] OBLOY E.J. and SHARETTS-SULLIVAN B.H., "Exploitation of Intellectual Property by Electronic Chartmakers: Liability, Retrenchment and a Proposal for Change", *Proceedings Conference on Law and Information Policy for Spatial Databases*, October 1995, Tempe (AZ), National Center for Geographic Information and Analysis (NCGIA), p 304–12.

[ONS 97] ONSRUD H.J., "Ethical Issues in the Use and Development of GIS", *Proceedings of GIS/LIS'97*, October 1997, ACSM/ASPRS, Cincinnati (OH), p 400–01.

[PAY 01] PAYETTE J., DESCHAMPS P., DESLAURIERS P., MASSE C. and SOLDEVILA A., *Responsabilité*, 2001, vol 1, Cowansville, Les Éditions Yvon Blais, p 190.

[PER 96] PERRITT H.H. Jr., *Law and the Information Superhighway, Privacy: Privacy, Access, Intellectual Property, Commerce, Liability*, 1996, Somerset, United States, John Wiley & Sons, p 730.

[PHI 99] PHILLIPS J.L., "Information Liability: the possible chilling effect of tort claims against producers of geographic information systems data", *Florida State University Law Review*, 1999, vol 26, number 3, p 743–81.

[REI 02] REINKE K. and HUNTER G.J., "A theory for communicating uncertainty in spatial databases", in *Spatial Data Quality*, 2002, SHI W., FISHER P. F. and GOODCHILD M. F. (eds), New York, Taylor & Francis, p 76–101.

[ROB 00] ROBERT P., *Le Nouveau Petit Robert, Dictionnaire Alphabétique et Analogique de la Langue Française*, 2000, Paris, Dictionnaires Le Robert, p 2841.

[ROU 91] ROUET P., *Les Données dans les Systèmes d'Information Géographique*, 1991, Paris, Éditions Hermès, p 278.

[VEZ 03] VEZINA N., "L'exonération fondée sur l'état des connaissances scientifiques et techniques, dite du 'risque de développement': regard sur un élément perturbateur dans le droit québécois de la responsabilité du fait des produits", *Mélanges Claude Masse*, LAFOND P.-C, 2003, Cowansville, Les Éditions Yvon Blais, p 433–66.

[VIV 02] VIVANT M. *et al.*, *Lamy Droit de l'Informatique et des Réseaux*, 2002, Paris, Lamy S.A., p 2002.

[WAL 02] WALFORD N., *Geographical Data, Characteristics and Sources*, 2002, London, John Wiley & Sons, p 274.

[ZHA 02] ZHANG J. and GOODCHILD M.F., *Uncertainty in Geographical Information*, 2002, New York, Taylor & Francis, p 266.

Appendix

Quality, and Poem Alike

Our European dictionaries are indebted to Cicero, a passionate expounder of Greek philosophy in loving command of his Latin language, for the word "*quality*". In creating the neologism *qualitas* from the pronoun *qualis* ("of what kind"), in his book *Academica* (45 BC) [CIC 51], he was importing the convenience of the word *poiotes* which Plato, in his *Theaetetus* some 320 years before, had invented from the pronoun *poios* ("of what sort") [PLA 26]: the convenience of designating directly, for an object, its state of being "as-such", of having the character it shows.

This meaning has barely survived, except sometimes in philosophy. We are close to this acceptation when the word is used as a synonym for "character" or "characteristic", when we speak of the quality of light or sound, or when subtlety and perfidy are ranked among qualities. Yet, carried along by the regular drift of centuries which transmutes differentiation into valuation, usage of "quality" is now structured in depth by scales of practical or moral values which serve to locate – and even interpret – the characteristics observed. More often than not, the word "quality" nowadays is prompted by, and conveys, the polarization of good and bad.

Cicero, in the same passage, is more specific about his coinage: "… the product of both force and matter, [the Greeks] called this … if I may use the term, 'quality' " (*Academica*, I.vi.). When it came to tracing the origin of quality, Antiquity was apparently satisfied with the conjugation of matter and force. Today's value-filters have made observers and users self-interested and demanding, and producers of goods or services cautious, and anxious to serve correctly. The quality of our time, as expressed, for example in the ISO 9000 standards, lies in the *management*, or,

---

Appendix written by Jean-François HANGOUËT.

more exactly, in the constant managing of the transformation of matter by force, at all stages of production – for the greatest satisfaction of all, undoubtedly.

Nevertheless, there is more to quality than meets today's eye. When brought side to side, the Greek terms *poiotes* (quality) and *poietes* (the *maker* of objects or odes, which to these days remains in our *poet*) look strangely alike. Gratuitous as it may seem, the play on Greek words however calls on Plato's kindly patronage; his taste for puns is well-known indeed, as reported by the Spanish philosopher José Ortega y Gasset, who, not without eagerness, provides additional evidence: "so Plato here and there in his last works enjoys playing with words, [for example] with these two terms which in Greek sound almost identically: παιδεία [*paideia*] – culture – and παιδιά [*paidia*] – childish play, joke, joviality" [ORT 60]

In all truth – which intuition presses to explore, once alerted by the play on words, and intrigued by the pair "made as such / make" which can be heard in the pair "quality / maker" – the formal similarity between *poiotes* (quality) and *poietes* (maker) is not fortuitous, but organic. Linguistics has shown that the two Greek terms share the same etymology. The keen Hellenist Pierre Chantraine [CHA 99] uses the exclamation mark to transcribe his surprise at the sudden resolution of a paradox by the obvious: "ποῖος '*of what sort-ness, of what nature?*' … provided the derivative ποιότης 'quality' coined by Pl., Tht 182a and, connected with ποιέω (!), the denomitave ποιόω [to make of such quality]" (*Dictionnaire étymologique de la langue grecque*, entry πο-). A few pages ahead, in the entry "ποιέω" (to make), comes his confirmation: "between ποιέω and -ποιός it is agreed that the radical for both words may be given as [Indo-European:] $^*k^w ei$- …, which suits both form and sense [the sense being that of *order*, of *setting in order*]."

Intuition can be pursued yet even further upstream, by looking at the entries for "*poésie*" ("poetry"), "*quel*" ("what") and "*qui*" ("who") in the *Dictionnaire historique de la langue française* [ROB 00] or in the *Oxford English Dictionary* [OXF 04]. No longer within ancient Greek, but catching today's *quality* of Latin ancestry, and *poem* from the Greek, an astonishingly common Indo-European background provides the basis for the two words: that of the etymons $°k^w o$-, $°k^w ei$-.

In the amazing perspective opened by the mutual presence that words have for each other, by their intimate kinship, possibly forgotten yet revealing in its turn when traced and found, a conclusion may be drawn: with *quality*, we are close to *poetry*.

Something we have always sensed, haven't we?

References

[CHA 99] CHANTRAINE P., *Dictionnaire étymologique de la langue grecque : Histoire des mots*, 1999, Paris, Klincksieck.

[CIC 51] CICERO, *Academica*, around 45BC. Edition consulted: *Cicero – De Natura Deorum; Academica*, translated by H. Rackham, 1933, revised 1951, Harvard, Loeb Classical Library, number, 268.

[ROB 00] *Dictionnaire historique de la langue française*, 1993 – 2000, Director Alain Rey, Paris, Dictionnaires Le Robert.

[ORT 60] ORTEGA Y., GASSET J., *What is Philosophy?* Edition consulted: *Qu'est-ce que la philosophie ?* Lesson VI, 1929. *Œuvres complètes 1*, translation by Yves Lorvellec and Christian Pierre, Paris, Klincksieck, 1988. The quotation comes from page 95. For an English edition: *What is Philosophy*, translation by Mildred Adams, 1960 New York, W.W. Norton & Company.

[OXF 04] Oxford English Dictionary (Second Edition) on CD-Rom, version 3.1, Oxford, Oxford University Press, 2004.

[PLA 26] PLATO, *Theaetetus*, around 369BC. Edition consulted: *Théétète*, translation by Auguste Diès, 1926, Paris, Les Belles Lettres.

List of Authors

Kate BEARD
Department of Spatial Information
Science & Engineering
University of Maine, Orono
USA

Jean BRODEUR
Centre for Topographic Information
of Sherbrooke (CTI-S)
Sherbrooke,
Canada

Nicholas CHRISMAN
Department of Geomatics Sciences
Laval University, Quebec
Canada

Alexis COMBER
Department of Geography
University of Leicester
UK

Sytze DE BRUIN
Department of Geo-Information
Science and Remote Sensing
Wageningen University
The Netherlands

Rodolphe DEVILLERS
Department of Geography & Centre
of Research in Geomatics (CRG)
Memorial University of
Newfoundland
Canada

Peter FISHER
Department of Information Science
City University, London
UK

Andrew FRANK
Institute of Geoinformation and
Cartography
Technical University of Vienna
Austria

Marc GERVAIS
Department of Geomatics Sciences &
Centre of Research in Geomatics
(CRG)
Laval University, Quebec
Canada

Jean-François HANGOUËT
Institut Géographique National (IGN)
Paris
France

Jenny HARDING
Ordnance Survey
Southampton
UK

Gary J. HUNTER
Department of Geomatics
University of Melbourne
Australia

Robert JEANSOULIN
LSIS, University of Provence
Marseille
France

Nicolas LESAGE
Institut Géographique National (IGN)
Paris
France

Thérèse LIBOUREL
LIRMM, University Montpellier II
Montpellier
France

Kim LOWELL
Department of Forestry & Centre of
Research in Geomatics (CRG)
Laval University
Canada

Daniel PILON
Centre for Topographic Information
of Sherbrooke (CTI-S)
Sherbrooke
Canada

Serge RIAZANOFF
Visio Terra
Paris
France

Richard SANTER
Laboratoire Interdisciplinaire en
Sciences de l'Environnement
Université du Littoral Côte d'Opale
France

Sylvie SERVIGNE
LIRIS,
INSA Lyon
France

Alfred STEIN
ITC
Enschede
The Netherlands

Sylvain VALLIÈRES
Centre for Topographic Information
of Sherbrooke (CTI-S)
Sherbrooke
Canada

Pepijn VAN OORT
Department of Geo-Information
Science and Remote Sensing
Wageningen University
The Netherlands

Bérengère VASSEUR
LSIS, University of Provence
Marseille
France

Richard WADSWORTH
Centre for Ecology and Hydrology
Monks Wood
UK

Index

A, B

abstract universe 27, 214
accuracy 23, 184
ambiguity 51, 130
atmospheric correction 86
attribute
   accuracy 149, 188
   uncertainty 93
Bayes 125
boundaries 96

C

calibration 84
capture specifications 152
CEN/TC 287 199
choropleth maps 89
civil
   law 283
   liability 285
classification 94
commission 189
communication 240
completeness 150, 183, 189
confidence 185
confusion matrix 230
connectivity 149
connotation 214
control data 37

currency 146

D

data
   provider 141
   quality 54, 240
   external 36, 39, 185, 255, 283
   quality visualisation 241
decision-making 271
Digital Elevation Model (DEM) 278
   accuracy 75
Dempster-Shafer 131
denotation 213
discord 52, 130

E

error 34, 48, 125
   model 110
   propagation 111
   -aware GIS 246
   -sensitive GIS 246
expert knowledge 130
external quality 255

F

FGDC 199
fiat 124